Discovering the
Nature
of Light

The Science
and
the Story

Discovering the

Nature

of Light

The Science
and
the Story

Norval Fortson
University of Washington, USA

NEW JERSEY · LONDON · SINGAPORE · BEIJING · SHANGHAI · HONG KONG · TAIPEI · CHENNAI · TOKYO

Published by

World Scientific Publishing Co. Pte. Ltd.

5 Toh Tuck Link, Singapore 596224

USA office: 27 Warren Street, Suite 401-402, Hackensack, NJ 07601

UK office: 57 Shelton Street, Covent Garden, London WC2H 9HE

Library of Congress Cataloging-in-Publication Data
Names: Fortson, E. Norval, author.
Title: Discovering the nature of light : the science and the story /
    Norval Fortson, University of Washington, USA.
Description: New Jersey : World Scientific, [2022] | Includes index.
Identifiers: LCCN 2021049417 | ISBN 9789811249594 (hardcover) |
    ISBN 9789811250293 (paperback) | ISBN 9789811249600 (ebook) |
    ISBN 9789811249617 (mobi)
Subjects: LCSH: Optics. | Light.
Classification: LCC QC350 .F67 2022 | DDC 535--dc23/eng/20211201
LC record available at https://lccn.loc.gov/2021049417

British Library Cataloguing-in-Publication Data
A catalogue record for this book is available from the British Library.

For any available supplementary material, please visit
https://www.worldscientific.com/worldscibooks/10.1142/12650#t=suppl

Desk Editor: Joseph Ang

Typeset by Stallion Press
Email: enquiries@stallionpress.com

Contents

Introduction

This book is a science text about light for the general reader; it is also an adventure story and a detective story. Readers can learn about the fascinating nature of light, share in the thrill of each discovery, find out who the discoverers were, and see how the pieces of the puzzle about light were deciphered and put together. There is another side to the book as well, namely the myriad applications that have been opened up by these discoveries, a subject that is so vast that we must be content with sketches of a few examples as we go along, such as fiber optics, the laser, and the recent optical detection of gravitational waves. I hope there are many readers who will enjoy all these features of the book.

The reader gets a sense of the spirit of the book right out of the starting gate in Chapter 1. We show how the French Philosopher and Mathematician René Descartes in the early 17th century used the Dutch surveyor Willebrord Snell's unpublished law of refraction to find the exact location of the rainbow in the sky. He calculated the bending of sunlight by refraction when it passes through raindrops, and as a result found the precise angle, 42°, that all rainbows are seen relative to the incident sunlight. This angle had been measured for centuries but not understood. With this celebrated triumph, Snell's law of refraction was accepted throughout Europe, in place of what had become dogma handed down from the Greeks. The modern age of optics and its applications was thus launched. There is an equally interesting story to be told about each subsequent advance. These advances at first proved conclusively that light is a wave, an electromagnetic wave in fact. Then at the beginning of the 20th century, a second set of advances proved equally conclusively that light is a beam of particles called photons. The book ends with the resolution of this wave–particle duality and a brief look at how light fits into the grander scheme of physical laws.

I hope the level of difficulty is about right for the majority of readers. Certainly the science and mathematics are too simple for physicists and many others, but they can enjoy the stories and may find many of them surprising. At the other extreme, some explanations will be too difficult for readers with little or no background in science, but they may still learn from the stories and gain some appreciation of how science works. Between these extremes, I hope there are many readers who find the book entertaining and enlightening and are motivated to concentrate hard on understanding the science.

Of course there is more to light than what is described in this book. Light is a source of beauty, expressed in paintings and photography, and an inspiration for literature and poetry, all beyond the realm of science. However, the science of light has its own beauty, all the varied properties of light fitting together in one theory, a beauty that I hope is conveyed in this book.

Acknowledgments

I thank my colleagues and students over the years for many discussions and insights about light. I also thank Albert Fuchs who read the manuscript and offered many valuable suggestions. I dedicate this book to my family, and especially to Alix Fortson, my wife of over 60 years, who saw me through yet another project with her usual spirit and enthusiasm.

Chapter 1

Reflection and Refraction of Light

"Considering that this [rain]bow appears not only in the sky, but also in the air near us, whenever there are drops of water illuminated by the sun, as we can see in certain fountains, I readily decided that it arose only from the way in which the rays of light [refract and reflect from] these drops and pass from them to our eyes. Further, knowing that the drops are round .... I had the idea of making a very large one, so that I could examine it better."

René Descartes, *Discours de la Méthode.*

In 1637, the famous French philosopher, mathematician, and scientist René Descartes startled the scientific world and attracted great public attention with his revolutionary theory of the rainbow and his remarkably successful calculations based on it. Not only did he show why sunlight is scattered so strongly at a single angle to create the concentrated light at the bow, but he was able to use his ideas to calculate the size of the bow in the sky shown in Fig. 1. Descartes found the angle, 42.5°, *exactly* the value it had been measured to be but not at all understood for centuries. In doing this, as we will show, he validated *Snell's Laws* of reflection and refraction which have been used to make great advances in optics ever since.

Of course, to most of us the most striking thing about the rainbow is the colors, but their existence in the rainbow was one mystery Descartes could not solve. That mystery was solved by Newton a half century later, as we will see in Chapter 3. When we think about it, however, a perhaps equally striking feature of a rainbow is that the light is concentrated in that narrow bow, and when we are told as shown in Fig. 1 that the light we see

1

Fig. 1. The angle made by each point of the rainbow with the sunbeam axis (as indicated here by the shadow thrown by the camera). Also shown are portions of the faint secondary rainbow above the primary. Derived from photo by Eric Rolph, via Wikimedia Commons. CC BY-SA 2.5, https://creativecommons.org/licenses/by-sa/2.5.

from the bow always makes the same angle of 42.5° with the axis of the sun rays, regardless of where it is viewed from or what the elevation of the Sun is, well, that is pretty amazing, too.[1]

Descartes found what happens when sunlight scatters from water drops in the sky, but to understand how he did it, we cannot start with that complicated problem; nor could Descartes! He started with the simpler situation of sunlight shining on some flat surface, say a quiet water pond, and undergoing reflection and refraction — a problem that had been studied for millennia. The correct paths of light after hitting such a surface had been found finally by Willebrord Snellius (Snell) in Holland in 1626, just 16 years before Descartes's great study. Snell had written down his laws of reflection and refraction but died in 1626 without publishing them, though many people were aware of them, including Descartes. In fact, it was only after Descartes had used these laws to get the exact angle of the rainbow that Snell's Laws were accepted, and from then on they have been

---

[1] Of course the colors broaden the rainbow a bit, and the angle of 42.5° is the angle measured to the red end of the spectrum of colors. From here on in this chapter we will approximate the angle as 42°.

used to make exciting advances right down to the present day, as in the fiber optic revolution. We ourselves will examine the paths of a sunbeam after hitting a quiet pond surface and learn what we will simply call *Snell's Laws*. Then we will be ready for Descartes to show that they are right.

Snell's Laws are the end result of familiar observations stretching back for thousands of years. We all have seen what happens when light encounters a flat surface of water such as a quiet pond. Some of the light reflects back into the air, some enters the water. Indeed, we know that the smooth surface of a quiet pond acts like a mirror since we can see the sky and trees reflected in it — and ourselves certainly if we lean out from the bank, just as according to the myth in which Narcissus saw his own image and found himself too entrancing. At the same time we know that some light enters the water instead and allows us to see objects below the surface of the pond, such as fishes or stones.

Thus, we can believe what is diagrammed in Fig. 2. When a beam of light in air arrives at a water surface, the beam splits into two beams, one

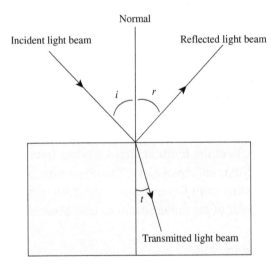

Fig. 2. A beam of light in air incident on a water surface splits into a reflected beam and a transmitted beam. The angle of reflection r measured relative to the normal to the surface is equal to the angle of incidence i. The angle t of the transmitted beam relative to the normal is smaller than the angle of incidence. This bending of the beam in transmission is called refraction.

reflected back into the air, the other transmitted into the water. If we measure the angles of these beams relative to the *normal* (the direction perpendicular to the surface of the water) as indicated in the figure, the angle r of the reflected beam is equal to the angle i of the incident beam, while the angle t of the transmitted beam is smaller. This bending of the beam upon entering (or leaving) the water is called *refraction*. The equality of the angles on reflection is already Snell's Law #1: it was known to the ancients. It's the law governing refraction that constitutes Snell's Law #2 that is hard, and makes the story interesting. As we might expect, a beam of light traveling upward in water bends away from the normal when it crosses the surface and enters the air above; in fact, the relation between the angles in air and in water is the same whether light enters or leaves the water. When we stand at the edge of a pond it is light leaving the water that we see when we spy an underwater fish; because of refraction, a swimming fish may not be where it appears to be, and from earliest times hunters spearing a fish must have learned to aim below its apparent position.[2]

Already in ancient times people had learned to go beyond such simple observations, and by the Hellenistic period the Greeks were making detailed measurements of refraction in water, glass, and other transparent materials. Such measurements, which marked the beginning of the scientific study of reflection and refraction, were repeated and recorded by Claudius Ptolemy around AD 150 in Alexandria. Beyond his being a quite famous Greco-Egyptian who held Roman citizenship, few reliable details of Ptolemy's life are available. He was best known for the *Almagest*, in which he described his geocentric model of the universe with all heavenly objects revolving about the Earth in circular orbits (more precisely orbits of orbits, called cycles and epicycles). This *Ptolemaic System* was a centerpiece of astronomy until Copernicus in the 15th century removed the Earth from the center of the universe and instead assumed it is merely one

---

[2] An interesting case arises when a beam "attempts" to cross a boundary into a medium of lower index than the one it is coming from, such as from water into air, when it can't do it. At angles greater than the so-called critical angle where the transmitted light is parallel to the surface, there is no place for the transmitted light to go, and then *all* of the beam is reflected back into the glass. This important effect is known as *total internal reflection* (TIR), and we return to it when we discuss fiber optics at the end of this chapter.

of the planets in orbit about the Sun. By then however, Ptolemy had become a legendary figure often called Ptolemy, King of Alexandria, and it took a fierce battle — political, religious, and scientific — to depose him and his system.

Of particular interest to us here, Ptolemy also had written a comparatively smaller work, *Optics*, about properties of light, including reflection, refraction, and color. Relatively small though it was, the work is of enormous significance for the early history of optics and was treated as a bible for 1,500 years. It contains the measurements on refraction mentioned above, with results tabulated systematically by angle. Ptolemy's measurements have been admired as the "most remarkable experimental research of antiquity" (George Sarton). However, though Ptolemy's tables of refraction were indeed based somewhat on careful measurements, unfortunately some values were adjusted just a little bit to conform to mathematical regularities that Ptolemy thought were indicated by the data. The prestige of the Greeks — and Ptolemy in particular — was so great that these adjusted values became part of the accepted wisdom of optics for some 1,500 years. Very little progress was made in the basic understanding of refraction for all this time because Ptolemy's reputation was so great that apparently when measurements disagreed with his, they simply were not published; they were thought to be wrong or at least too risky for the reputation of the scientist.

Part of Ptolemy's table of refraction from air to water is shown in Table 1, based on experiments similar to the one in Fig. 2. As explained in the caption to the table, a pattern in the data is clear that we now know is wrong. That was the adjusting that we were talking about. As a result, there were small errors that were still large enough that the discovery of the correct laws of refraction was delayed by centuries. In the table we also show in the third column the refraction angles we now know are correct and in the rightmost column we show the ratio of the trigonometric sines of these same angles; note this ratio is a constant, the same for all angles! Its value is called the *index of refraction* of water relative to air.

In general, a similar relationship holds when light crosses the boundary between any two media. It is Snell's Law, written as:

$$\sin \theta_1 / \sin \theta_2 = n_2/n_1 \qquad (1)$$

Table 1.   Showing for each incident angle i, the transmission angle t_p in Ptolemy's table, and the actual transmission angle t as measured today. The latter angles yield a constant value for sin i/sin t, which is an expression of Snell's Law. The constant is the index of refraction of water relative to air. Note the pattern of the values of t_p in Ptolemy's table, the difference decreasing by 0.5° for each successive 10° in i, a pattern he thought he saw and hence imposed in the data but is clearly an error when compared with values today. These erroneous values held back progress for 1,500 years.

Refraction			
Ptolemy		**Actual**	
i	t_P	t	Sin i/Sin t
10	8	7.50	1.33
20	15.5	14.90	1.33
30	22.5	22.08	1.33
40	29	28.90	1.33
50	35	35.17	1.33
60	40.5	40.63	1.33
70	45.5	44.95	1.33
80	50	47.77	1.33

Note: The sine of an angle is just a short hand for writing the opposite side divided by the hypotenuse of a right triangle formed by the angle.

where 1 and 2 refer to the two media, for example water and air, while θ_1 and θ_2 are the angles in each medium that the light beam makes with the normal at the boundary. It is found that the same value of n works for a material regardless of what material it is paired with. Therefore, we can speak of **the** index of the material, and the right side of Eq. (1) is the ratio of the indices of the two media.[3] In the case of glass, the index is found to be 1.5, so we write for the air–glass interface sin θ_A/sin θ_G = *1.5* and for water–glass, sin θ_W/sin θ_G = 1.5/1.3 = 1.2.

---

[3]When the light beam is incident from the higher index medium interesting things can happen, such as total internal reflection at the boundary. We will discuss this effect at the end of this chapter, in connection with fiber optics.

The Greeks, including Ptolemy, never discovered Eq. (1). In 1621, this simple law was written down by Willebrord Snellius (Snell) of Holland, and came to be called Snell's Law of Refraction. We don't know how Snell came upon this relationship. There is no record of him having done accurate enough experiments to convince himself that Ptolemy was wrong. We do know that Snell was expert in surveying and quite familiar with the use of trigonometry in it. That somehow may have led him to the law. But he did not publish anything; in fact he simply put the statement of his law in a drawer. There is an interesting story here, but we just don't know it. But we now know it is correct.

Even after Snell, since he didn't publish his laws, they might have lain longer in obscurity *except for Descartes*. Descartes burst upon the scene and upset the status quo in just about everything he touched. The theory of the rainbow is no exception. Descartes had access to Snell's unpublished laws, had the instinct to appreciate their mathematical simplicity, and in a remarkable calculation he discovered that they predicted the size of the rainbow exactly. After Descartes published his results, Snell's Laws were believed almost immediately and modern optics was launched. This is the central story of this chapter.

Descartes is often considered the father of Western philosophy, though from the beginning his views inspired heated disputes. Descartes rejected the authority of doctrines and famous thinkers and indeed of his own senses. Instead, he became popularly known for his procedure of *systematic doubt* leading to his famous assertion *cogito ergo sum — I think therefore I am.* Many have argued that this statement shifts the authority for truth from God to people, and from what is decreed as "true in church doctrine" to "of what can **I** be certain", and thus is a crucial part of the change from the Christian medieval period to the Enlightenment.

In mathematics, Descartes is credited with founding analytical geometry by introducing his stratagem of specifying the position of a point on a plane by its distance from two fixed lines or coordinate axes. This *Cartesian coordinate system* named after him allows algebraic equations to be expressed as geometric curves and vice versa.

In science, Descartes advocated the use of mathematics, particularly in physics as in his work on the rainbow. Bertrand Russell wrote of Descartes: "In philosophy and mathematics, his work was of supreme importance; in science, though creditable, it was not so good as that of some of his contemporaries." If we note that these contemporaries in science included Galileo Galilei and Christiaan Huygens, Russell's evaluation could be accepted and still allow us to rank Descartes highly in science. Also, we are not concerned with his full body of work, but instead we want to particularly honor his explanation of the rainbow as a brilliant advance and a crucial part of our story.

Descartes was born in La Haye en Touraine (now named Descartes) in France, on 31 March 1596. When he was one year old, his mother died and René lived mainly with his grandmother and great-uncle. Descartes's father, a member of the minor aristocracy, had little admiration for Descartes or his achievements, and is reported to have said on the publication of Descartes's first and perhaps most influential book, the *Discourse on the Method* and accompanying essays, in 1637:

> "Only one of my children has displeased me. How can I have engendered a son stupid enough to have had himself bound in calf."

Despite fragile health, Descartes entered the Jesuit College at La Flèche, a school for brilliant students. Then, at the University of Poitiers, Descartes half-heartedly followed his father's wishes and earned a degree in Canon and Civil Law. He then moved to Paris where in his *Discourse on the Method* he recalls:

> "I entirely abandoned the study of letters. Resolving to seek no knowledge other than that of which could be found in myself or else in the great book of the world, I spent the rest of my youth traveling, visiting courts and armies, mixing with people of diverse temperaments and ranks, gathering various experiences, testing myself in the situations which fortune offered me, and at all times reflecting upon whatever came my way so as to derive some profit from it."

Descartes had aspired to be a military officer, and in 1618 he joined the Dutch States Army in Breda where he undertook a formal study of military engineering and was encouraged to advance his knowledge of mathematics. There he met and studied with Isaac Beeckman, a leading expert on the uses of mathematics in the study of motion. Descartes left

Fig. 3. Portrait of René Descartes by Frans Hals.

the army after two years and then sold all of his property to invest in bonds, which provided a comfortable income for the rest of his life.

Descartes returned to the Dutch Republic in 1628, a protestant country with a tradition of tolerance, where Descartes — a Catholic — wrote all his major work in philosophy, mathematics and science. Some of the time he stayed with Christiaan Huygens' family and had a major influence on Christiaan's basic scientific views, as we will see when we look closely at the latter's life in the next chapter. In Amsterdam while teaching at Utrecht University, Descartes had an intimate relationship with a servant girl Helena Jans van der Strom, with whom he had a daughter, Francine, who died at age 5 in 1640, a major personal sorrow for Descartes.

Descartes was considered a difficult person — arrogant and self-centered. He even accused his helpful senior colleague and teacher, Isaac Beeckman, of plagiarism, and some of Descartes's low personal reputation was clearly deserved. Still, much also was due to false rumors by vindictive enemies in an age of accusation and rivalry. Such vindictive rivalry probably underlay the well-known stories about Descartes's appropriating the unpublished Snell Laws as his own. He certainly put them to use however.

The case of the rainbow stands apart. Descartes's method of finding the solution was original as were his calculations that yielded striking

Fig. 4. *The Rainbow Landscape* by Peter Paul Rubens, 1636. The primary and secondary rainbows are each shown (albeit without the full spectrum of colors).

results, and he received public recognition for these genuine revolutionary achievements from the beginning.

In Descartes's time we can imagine that most people were fascinated by the rainbow just as they are today. Painters were no exception. In 1636, the year Descartes began studying the rainbow, Peter Paul Rubens painted *The Rainbow Landscape* shown in Fig. 4, a bucolic scene viewed from Rubens' house, dramatically framed by a bright rainbow with part of a secondary rainbow above the primary.[4]

The rainbow is produced by viewing sunlight that has scattered from countless raindrops. In Fig. 5(a) we show a ray of sunlight deflected by a

---

[4]Actually, a rainbow is not a fixed part of the landscape as a tree is, but will occur in a different place for each person in the painting. So the question is, whose rainbow did Rubens paint? Probably the painter's (his own) view. No one else is even looking at the rainbow; an enlarged view confirms that the wagoner on the far left is actually looking at the young woman in red coming in from the field, who is returning his gaze intently.

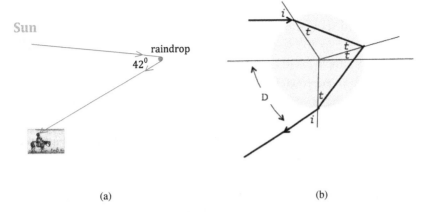

(a)   (b)

Fig. 5.   (a) Sideview of light path for just one of countless raindrops creating the rainbow; (b) Enlarged view of the raindrop, showing the Descartes Ray for which *i* is about 58° and *D* is about 42°. All other rays have a smaller deflection *D*, whether the incidence angle *i* is smaller or larger than 58°.

particular one of those drops and then viewed by an observer. Descartes decided to investigate the paths the sunlight follows in such an individual raindrop. As indicated in the epigraph to this chapter, he filled a spherical flask with water to serve as an enlarged model of a raindrop, and homed in on the key trajectory by observing the trajectories of a light beam through it.[5]

Descartes shone a bright horizontal sunbeam through his globe. Each part of the sunbeam hitting the drop follows a different trajectory in the drop depending upon where it hits the surface of the drop and the refractive path it then follows. Each one of these trajectories we call a *ray*. He noticed a colored spot appeared on the lower surface of the globe, and quickly figured out that spot was where the sunbeam inside the globe was emerging after reflecting off the back of the globe, forming what is now

---

[5]The thickness of the glass wall forming Descartes's globe, even if narrow, does of course affect the beams passing through it. However, the bending of the beam entering the wall is compensated well by the opposite bending leaving the wall *except for the water*, so that the net effect is close to that of the water alone.

called the Descartes Ray in Fig. 5(b). Descartes found the deflection angle of this colored spot, shown in Fig. 5(b) as the angle D, was close to 42°, about the same as the well-known rainbow angle shown back in Fig. 1! (It is a fairly simple exercise in geometry of a circle to show that all internal angles in Fig. 4b are equal to t, the initial transmission angle, and $D = 4t - 2i$.) Descartes could not explain the colors of the rainbow; nor could anyone until Newton 35 years later. He did know, however, that whatever causes the color, the ray he had identified must be the key to understanding at what angle the rainbow appears in the sky.

At this point he had not gone much further than others before him[6] who had built globes of water and observed colored rays at the rainbow angle of 42°. In a sense he and they had merely proved that a spherical globe of water deflects light the same way as individual raindrops do. Descartes wanted to explain or understand why this ray at 42° was so singular, why so much of the sunlight that enters the globe (or individual raindrops) is deflected near it; in other words still, why does the rainbow exist? Somehow the laws of reflection and refraction should explain why most of the deflected light is near that one angle. He says:

> "I took my pen and made an accurate calculation of the paths of the rays which fall on the different points of a globe of water to determine at what angles, after 2 refractions and one reflection [using Snell's Law with the refractive index of water] they will come to the eye, and then I found that there are many more which can be seen at an angle of 42° than at any smaller angle and that there are <u>none</u> which can be seen at a larger angle."

Descartes's results are plotted in Fig. 6. Descartes figured out why just rays around one trajectory dominate the view. It is because it is the maximum deflection as shown by his calculation; rays with larger angles of incidence are deflected at a smaller angle again. It is trajectories clustered near this maximum that far outnumber all others. As the caption to Fig. 6 points out, the deflection angle differs by less than 2° from its maximum of 42° as the angle of incidence i varies over a wide range, from 50° to 70°. Therefore, these rays dominate the light scattered from the drop, and

---

[6]In the 14th century both al-Fārisī of Persia and Theodoric of Freiberg had built large models of a spherical raindrop. However, this fact was unknown to Descartes.

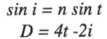

$$\sin i = n \sin t$$
$$D = 4t - 2i$$

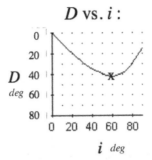

D vs. i :

i *deg*

Fig. 6. The results of Descartes's calculation of the deflection angle D defined in Fig. 5 for each angle of incidence i using Snell's Law. Note that the rays cluster about the maximum $D = 42°$ to form the rainbow, differing by less than $2°$ in D from $i = 50°–70°$.

account for the location of the primary rainbow. Small wonder that most of the light is seen near this angle! This clustering near an extreme value is characteristic. For example, a ball tossed 20 feet straight up spends far more time in the top vertical foot of its trip than in any other vertical foot of its path.

This special angle *found with Snell's Law and the index of refraction of water* turned out to be exactly the angle measured for centuries for the rainbow. Snell's Laws account for the rainbow!! Also, through these simple but profound statements the fundamental problem of the size of the rainbow was solved for the first time.

In fact, we often see *two* rainbows; the so-called secondary one, usually much fainter, is larger than the primary one but with the same center as shown in Fig. 7. Well, Descartes found from his theory that there *should* be a second one, with rays taking one more bounce inside each droplet than for the single rainbow. The angle he predicted was precisely the angle that is observed, so he got <u>both</u> rainbows right.

What about Ptolemy's measurements shown back in Table 1? Using them with Descartes's method instead of using Snell's Law did not yield the correct primary rainbow angle, but the disagreement was fairly small. However, Ptolemy's measurements did not yield at all the correct secondary rainbow, so they had to be discarded at last.

Fig. 7.   A double rainbow. The primary is brighter and has the radius of 42° already discussed. The secondary has the colors reversed and has a radius of 51°. Descartes correctly calculated both angles using Snell's Laws. Photo by Lars Leber Photography. July 16, 2019.

If you were not already convinced that Descartes knew what he was doing, you must be by now. His contemporaries certainly were, and this work became celebrated across Europe and with it the belief in Snell's Laws. Before Descartes, refraction experiments were still inconclusive. Descartes showed that Nature itself had done the great experiment and had presented the results in the sky — the rainbow — which was nearly indisputable evidence that Snell's Laws are correct. The primary and secondary rainbows together provided an exacting test for all to see.

It took Descartes just one or two sentences to state his conclusion. It took laborious calculations "after he took his pen" to reach it. Indeed, Descartes had brilliant insights about the rainbow, which led him far, but he had the final and truly original insight only after persistence led him to it. That is part of why this is such a wonderful story, because Descartes is one who cultivated the reputation of being quick and brilliant, and indeed he was, but his final great conclusion came to him only after determined and surely tedious work. And remember, he could not have been certain; Snell's Law was not established knowledge yet — it wasn't even published — so Descartes was putting massive effort on the conjecture that the law was

A Recently Uncovered Surprise

Before turning to the application of Snell's Laws and their significance, let us go back for a startling historical fact that was uncovered in the 1990s, namely that Snell's Law of refraction actually had been discovered 600 years *before* Snell, by Ibn Sahl in Baghdad during the *Islamic Golden Age*, which lasted from the 8th century to the 12th century. The discovery was lost when Baghdad was completely destroyed by the Mongols in the1200s and so the world waited hundreds of years for the progress from Snell's Law to begin.

This age began with the House of Wisdom in Baghdad, where scholars from different parts of the Islamic world were required to come and translate all knowledge into Arabic. Scholars had great prestige during the Golden Age, with salaries estimated *to be the equivalent of professional athletes today*. Science, culture, and the economy prospered.

A chief interest during the Golden Age was to start fires by focusing sunlight. A familiar modern example is a pocket magnifying glass used to ignite bits of paper. Ibn Sahl developed his version of the law of refraction in the course of writing his treatise: *On Burning Mirrors and Lenses*. (We will be discussing a part of this treatise that was reconstructed by Roshdi Rashed from two fragments in two museums and edited in 1993.) A burning lens is a large convex lens that can concentrate the Sun's rays onto a small area, heating up the area and thus resulting in ignition of the exposed surface. Burning mirrors achieve a similar effect by using reflecting surfaces to focus the light. The Greeks used concentrated sunlight for lighting practical as well as ceremonial fires, and also widely to cauterize wounds. A dramatic goal was inspired by a legend about Archimedes, who was said to have used a burning glass or mirror (or several of them) as a weapon in 212 BC, when Syracuse was besieged by Marcus Claudius Marcellus. The Roman fleet was supposedly incinerated, though eventually the city was taken and Archimedes was slain. Although such destructive weapons seem not plausible by modern estimates, the legend of Archimedes gave rise to active research on burning glasses and lenses until the late 17th century.

Ibn Sahl showed that a reflecting surface in the shape of a parabola made a perfect mirror, that is it focused the Sun taken as a point source of light to a single point. What was revolutionary was his calculation with *refracting lenses*. So far as we know, Ibn Sahl was the only one who actually

(Continued)

(Continued)

found the exact law of refraction to do the calculations with lenses, and this is the part of his work that was only discovered in the 1990s. His method of stating the law was geometrical but is completely equivalent to Snell's Law and is illustrated in the following figure.

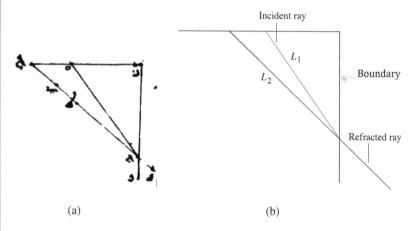

(a)   (b)

(a) Ibn Sahl's correct geometrical method of finding the law of refraction 600 years before Snell. (b) Drawing based on original: $L_1/L_2 = n_1/n_2 =$ const for all angles. The interested reader can verify this relation is equivalent to Snell's Law of refraction. Vertical boundary between two media is shown.

Source: Wiki File:Ibn_Sahl.svg.

He knew the measurements of Ptolemy, but as we have commented above, Ptolemy had adjusted some values which would not agree with the correct law. Perhaps Ibn Sahl had access to more reliable values which he could use to find the general, relatively simple rule; why he could do it and no one else did for over 600 years is a mystery, or a testament to his genius. He also found that a lens in the shape of a hyperbola is a perfect lens, focusing to a single point. This finding of Ibn Sahl is further proof that he had the correct law of refraction since we can show that this shape focuses to a point using Snell's Law.

So, what happened to all the knowledge about optics from the Golden Age centered in Baghdad? We have no idea what else was known and how it would have been preserved if the Golden Age had continued and the city had remained one of the most brilliant intellectual centers in the world. In

(Continued)

fact, though, all knowledge in Baghdad perished and the scientific spark of the Golden Age seems to have perished with it when the city itself was completely destroyed and its libraries razed in 1258 AD by the Mongols. The general consensus has been that the Golden Age elsewhere ended with this destruction.

The Mongol commander Hulagu, having decided to conquer both Baghdad and Persia, assembled the most numerous Mongol army to have existed, 150,000 strong, supplemented by large Christian forces plus about 1,000 Chinese artillery experts. The Caliph Al-Musta'sim refused Mongol demands for tribute, and although he had failed to prepare for an invasion, the Caliph believed that Baghdad could never fall to invading forces and refused to surrender. Employing siege engines and catapults, the Mongols and allies breached the city's walls, and then looted and destroyed mosques, palaces, libraries, and hospitals, Estimates of the death toll range up to a million. Of course, the Grand Library of Baghdad (the House of Wisdom), which contained countless precious historical documents and books on subjects ranging from medicine to astronomy, was destroyed. Survivors said that the waters of the Tigris ran red with blood and black with ink. Clearly, no burning lens weapons appeared to save the city by incinerating the invading ships on the Tigris, although burning lenses and mirrors may well have been used to cauterize wounds during the sieges. The Islamic Golden Age came to an abrupt end, and it was centuries before Baghdad became important again.

Why didn't the discovery of Snell's Laws by Ibn Sahl lead to an equally early advance in its applications? To start with, he didn't have his Descartes! Maybe al-Fārisī of Persia or Theodoric of Freiberg who built models of the rainbow in the 14th century would have had Descartes's ingenuity, but by that time no one knew about the laws discovered by Ibn Sahl! Most of that had been destroyed by the Mongols by then. That is not to say by any means that all Arab knowledge perished. The knowledge from Baghdad had spread to many other places in the Muslim world and elsewhere before Baghdad was destroyed; for instance, the two fragments of Ibn Sahl's manuscript were found in two museums. As another example, it has long been known that Alhazen, in about 1000 AD and following Ibn Sahl, wrote the treatise *Book of Optics*, which later greatly influenced Europeans in the Middle Ages.

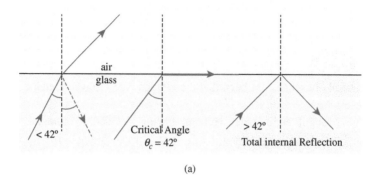

(a)

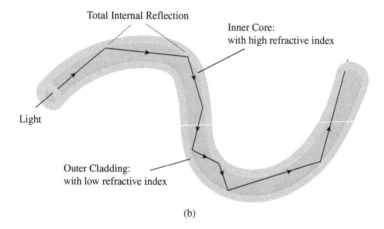

(b)

Fig. 8. (a) Total internal reflection (TIR) at glass-to-air boundary. (b) Light travels tremendous distances by total internal reflection in optical fibers.

Note: This illustration in Figure 8(b) is indicative but oversimplified. The thickness of the cables is comparable to the wavelength of light, which requires the full wave theory to treat the problem correctly, and we are not ready for that yet.

right. At one point or more during this process, Descartes must have felt exhilaration, as his purely theoretical analysis was leading to an exact number often measured but never explained until that moment.

Descartes's calculation of the rainbow amounted to a public demonstration of the correctness of Snell's Laws. Such accurate comparison between theory and experiment was unprecedented; the calculation itself was complicated and still the agreement was good to 2%! Anyone could imagine sunlight would be scattered by raindrops, but why so strongly at this *one particular* angle? Descartes, using Snell's Laws, told the scientific world why.

To return to our main story, almost immediately after Descartes's work on the rainbow verified Snell's Laws, attempts began to understand *why* light obeys Snell's Laws. The answer to the question "what is light?" must also answer why Snell's Laws work. Christiaan Huygens gave his answer as told in the next chapter, which is the answer we believe today, that light is a wave. But it wasn't the only answer that people believed at the time, as we will see.

With Descartes's success on the rainbow as inspiration, Snell's Laws were used to design better lenses of all kinds for eyeglasses, telescopes, and microscopes. Although slow at the beginning, the pace of advances based on Snell's Laws picked up steadily and advances have continued ever since, today including the fiber optics revolution that is so well known in medicine and in long-range communications. Fiber optic cables are a most spectacular use of *total internal reflection* (TIR), a phenomenon which was mentioned in the footnote below Eq. (1). TIR arises when a beam "attempts" to cross a boundary into a medium of lower index than the one it is coming from, such as from glass into air as shown in Fig. 8(a). The so-called critical angle occurs where the transmitted light is parallel to the surface. Using the notation below Eq. (1), this time for glass-to-air, this critical angle in glass is given by: $\sin \theta_G = 1/1.5$ or $\theta_G = 42°$. At incident angles greater than the critical angle, there is no place for the transmitted light to go, and then *all* of the beam is reflected back into the glass. Therefore, as illustrated in Fig. 8(b), the light that carries signals for computers and other electronics can be trapped inside tiny cables by TIR and piped through them for both short and very long distances.

What about Descartes himself? In the late 1640s Descartes had close intellectual associations with two learned ladies, first with Princess Elisabeth of Bohemia and later with Queen Christina of Sweden. He exchanged many letters with Princess Elisabeth about morality and psychology, which led to his treatise *Passions of the Soul*. Meanwhile, Queen Christina invited Descartes to her court in 1649 to head a new scientific academy and as well to tutor her on his ideas about love. She read his *Passions of the Soul*, which Descartes had sent her, and urged Descartes to publish it, which he did and dedicated it naturally to Princess Elisabeth. Descartes, apparently reluctantly, accepted the invitation to Christina's court, and she dispatched a Swedish warship to fetch him. She wanted daily lessons from him, but could spare the time only at five in the morning. Perhaps it was the early hours, the Swedish winter, or Descartes's delicate health, but he fell ill and died in February 1650. In the official story he came down with a cold on February 1, which became pneumonia and he died on February 11. However, in death as in life Descartes seems to have been the center of dispute, with conflicting claims by physicians at court and now recent books keeping the disputes alive, including the theory he was assassinated.

Whatever the truth of these claims, the positive accomplishments of Descartes in philosophy and mathematics were revolutionary, and for our story his work on the rainbow set optics on the correct course at last. We now proceed to continue on this course in the following chapters.

Chapter 2

Huygens Founds the Wave Theory of Light

"I have now shown how we may consider light as propagated, in time, by spherical waves ... I call these waves because of their resemblance to those which are formed when one throws a pebble into water and which represent gradual propagation in circles, although [the latter are] confined to a plane surface."

Christiaan Huygens, *Treatise on Light* (Paris 1678, Leyden 1690).

Christiaan Huygens was born in 1629, two years after Snell had died with his laws of reflection and refraction unpublished and 7 years before Descartes used Snell's laws to solve the problem of the rainbow which verified these laws and started the field of modern optics. As a child Christiaan marveled at water waves and how they propagate and as an adult he came up with a theory of how all waves propagate and applied it to light. This theory, called Huygens' Principle, is used to this day to describe light waves. Huygens' wave theory gave a natural explanation of reflection and refraction, but in addition it explained just as naturally the *exact form of Snell's Law itself*. This was a great triumph of the wave theory. However, although Huygens was celebrated in his lifetime for this and many other achievements, his wave theory was largely forgotten after his death in favor of Isaac Newton's particle theory of light, mainly because Newton was revered so much for his great law of gravity and laws of motion and, as we will see in Chapter 3, his discoveries about color. It wasn't until after 1800, when Thomas Young and Augustin Fresnel put forth their wave theories of light that were built on Huygens' ideas of a century earlier, that Huygens' wave theory was revived and eventually triumphed. But that is getting ahead of our story. In this chapter, we will see how Huygens developed his wave theory of light.

Huygens came from a cultivated and prominent family in The Hague, Netherlands. His father, Constantijn, was an important diplomat and statesman, also well known as a Latinist, poet, musician, and painter. While a diplomat in England, he became a good friend of John Donne, played the lute at the court of James I, and received an English knighthood in 1622. He was an advisor to the royal court at The Hague and had a major influence on cultural development in the Netherlands. He visited Leyden where he discovered Rembrandt van Rijn in the year of Christiaan's birth, and brought the young painter to the attention of Frederik Hendrik, the Prince of Orange and thereby procured for Rembrandt important commissions from the court.[1] The elder Huygens also knew Vermeer in nearby Delft, and likely introduced him into elite society. This same Constantijn Huygens corresponded for years with René Descartes, who visited the Huygens home frequently when Christiaan was young and was much impressed with his great skill in geometry. Descartes's ideas made a lasting impression on young Christiaan, especially the belief that any action such as transmission of light or the force of gravity between two distant objects must be conveyed by some subtle medium between them. Christiaan, his elder brother (named Constantijn after their father), and the two younger brothers were taught science and mathematics by a private tutor, as well as singing, playing the lute, and composition of Latin verses. Christiaan particularly enjoyed mathematics, drawing, and making mechanical models. Later, Constantijn would work closely with Christiaan on building telescopes.

Christiaan was somewhat delicate as a child, and did not exhibit the physical vigor of his lusty father as he grew up, but his mental energy came to match or surpass that of anyone he knew. At age 16, Christiaan entered the University at Leyden and two years later the new (and

---

[1] There are several letters from Rembrandt to *my most gracious lord Huygens*, seen as his protector, about progress on his paintings and compensation for them. These are the only letters still extant from Rembrandt to anyone.

short-lived) College of Orange at Breda (which he left somewhat early, apparently after a younger brother Lodewijk engaged in a swordfight with another student). Christiaan then accompanied his father's high-ranking friend, the Count Hendrik of Nassau, on a diplomatic mission to the splendid court of Frederik III, the new King of Denmark. Christiaan found it a court of pageantry: theater, expensive attire, and especially music, dancing, and ballet. During his stay he did what according to him in a letter to his brother Constantijn were the only occupations there: "eating and drinking, dancing and playing." After dinner "we relaxed *ins frauenzimmer* where there were 12 ladies of the Queen and some Fräuleins, all dressed in French fashion but not a single one speaking French." His father had in mind many such visits among the courts and elite society of Europe to promote his son in a diplomatic career like his own, but he soon realized that Christiaan's real passions were science and mathematics. Furthermore, he was truly brilliant at them. The elder Huygens also realized that this brilliance would shine in the courts of Europe, and used his remarkable connections to gain access for his son, which eventually helped lead to Christiaan's great success at Louis XIV's court at Versailles and his preeminent position in the French Academy of Sciences.

Thus, very early Christiaan (hereafter we will call him Huygens) began working on a wide range of problems in mathematics, physics, and astronomy, and by age 22 was already establishing a solid reputation in mathematics. He started constructing his own telescopes, with the strong encouragement of his father and skillful collaboration of his brother Constantijn.[2] In 1655, at age 26 he used their 12-foot telescope to discover Saturn's moon Titan, the first new moon observed since Galileo had seen Jupiter's moons 41 years earlier. He also saw the peculiar shape of Saturn that Galileo had described as a central sphere flanked by two adjoining

---

[2]Renowned lens makers usually guarded their methods with absolute secrecy. The Huygens brothers, in contrast, published their methods in Robert Smith's *Opticks* in England. They had great success using very strong pressure to polish lenses. They placed the lens on a table with the template on top of it and emery between the surfaces, and then slid the template back and forth sideways while a strong vertical force pressed the surfaces together. This force was created by the intriguing method of pushing down on the top piece with a long vertical pole wedged between it and the ceiling and bent like a drawn bow, which could create a large downward pressure while allowing freedom of movement.

smaller spheres, which after 40 years continued to puzzle astronomers. Huygens then cleared up the mystery the following year after he moved to Paris with his father and family. There a new 23-foot telescope was quickly assembled, and with that instrument Huygens concluded that the peculiar features were in fact one thin circle of matter around Saturn, and his sketch of *Saturn's ring,* as we now call it, was widely publicized. Huygens became famous.

Huygens had wide-ranging interests and made ground-breaking innovations in a number of fields.[3] He constructed the first pendulum clock, a revolutionary development in timekeeping. To make better clocks he studied the motion of a pendulum very carefully, and obtained the relation between the period of oscillation of a pendulum and the acceleration of gravity, g. He used the relation to measure g much more simply and with better accuracy than by the method of timing free fall in a vacuum. To solve the pendulum problem he had needed to understand its circular motion, and he was the first to analyze such motion correctly.

Huygens (Fig. 1) became a member of the French Academy of Sciences, having been recruited by Louis XIV's finance minister, Colbert, to be a founding member in 1665 with a fine apartment as the most prestigious member. During much of his residence in Paris during the 1660s and 1670s, France and Holland were at war, which is rarely mentioned in Huygens' correspondence. In that era, however, it was not unusual for someone of Huygens' station to live in an "enemy" country.

He also collaborated with the Danish scientist Ole Rømer (whom he knew in Paris and who played a key role in the first measurement of the speed of light, as we will see in Chapter 4) and fellow Dutchman Nicolaas Hartsoeker to design an efficient microscope for studying the new world of microorganisms just opening up in the 1670s with the Dutchman van Leeuwenhoek's pioneering studies.[4] Interest in this field took off, and the

---

[3] He enjoyed composing songs with harpsichord accompaniment, an instrument he apparently played well. He became intrigued by how to tune the harpsichord to achieve the best *temperament*; for example, equal temperament is a compromise that divides the octave into 12 equal semi-tones. This problem is one to which he applied his considerable mathematical abilities and his good musical ear.

[4] Van Leeuwenhoek was from Delft near the Huygens family home, and interest in him became a Huygens family affair. The father Constantijn knew van Leeuwenhoek, wrote

Fig. 1.   Christiaan Huygens, Pastel Portrait by Bernard Vaillant.

Paris microscope of Huygens, Hartsoeker, and Rømer, which enabled rapid frame-by-frame comparison of specimens mounted on a wheel, was quickly put into widespread use.

Huygens' greatest achievement, however, was his monumental wave theory of light, developed and first published in Paris. He endeavored to explain the most common observations of light in terms of light waves, quite successfully in the judgment of posterity but at odds with the particle theory of light that prevailed at the time. Huygens himself was intrigued by wave motion from early days in Holland, and later described how great waves of swaying grain would sweep across large fields near his birthplace. Today too, wave motion in one form or another intrigues most people. Ocean waves at the beach are ever popular, huge tidal waves are fascinating — and frightening — and nowadays "The Wave" at sports stadiums is an exciting way to root for the home team or to express spirit even at Wimbledon with members of the royal family doing their part. In Fig. 2, we show examples of many types of wave pulses: one-dimensional

---

poems about the microorganisms, and with his connections assisted the physician Reignier de Graaf to introduce van Leeuwenhoek to the Royal Society in London, which published all of his scientific work.

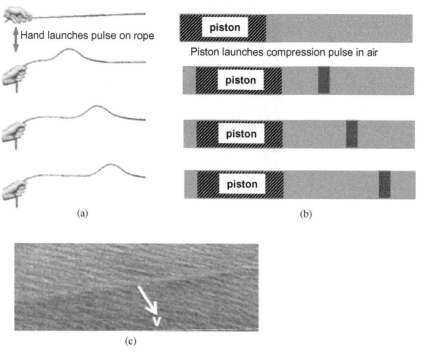

Fig. 2. (a) A transverse wave pulse on a rope, activated by hand, showing the successive positions of the pulse at later times. The rope itself moves up and down as the pulse passes; this disturbance of the rope, not the rope itself, moves to the right. (b) A longitudinal sound wave pulse in a hollow pipe, activated by a piston, showing the successive positions of the pulse at later times. This disturbance of the air, not the air itself, moves as the pulse to the right. (c) A water surface wave. As in (a), a transverse wave, but on water. Again, the water moves up and down as a wave disturbance passes and it's the disturbance (not the water) that moves in the direction of the wave velocity **v**.

(1D) waves on a rope in Fig. 2(a) or in a sound pipe in Fig. 2(b), and 2D waves on the surface of water in Fig. 2(c).

As emphasized in the Fig. 2 caption, all these waves exhibit a key feature of wave motion: a wave is a moving *disturbance* of a medium, not forward motion of the medium itself. "The Wave" in a stadium illustrates this essence of wave motion very well, since it is a traveling *disturbance* (people standing or sitting) of a medium (the crowd) rather than forward motion of the medium itself (the people are not moving around the stadium, their standing and sitting is). Leonardo da Vinci vividly captured this same essence of waves back in the 15th century: "...the [water] wave flees the place of its creation, while the water does not; like the waves made in a field of grain by the wind, where we see the waves running across the field while the grain remains in its place." Likewise a tidal wave is a traveling disturbance (rising and falling) of the water rather than actual water moving forward at the break-neck speed of the disturbance. Sonic booms are similar examples, in this case sound in the air. The air itself doesn't move forward at the speed of sound, the disturbance moves.

What about light waves? Again, it is the disturbance that moves but in what medium? Huygens postulated the luminiferous aether, a subtle medium pervading all space but not hindering the motion of physical objects. (Clearly much more subtle than air!) That concept posed fundamental conceptual difficulties, and in fact went out of favor for a century while Newton's particle theory held sway, but came back and reigned supreme in the 1800s until it was superseded by the more satisfactory idea of *fields* (not grain fields of course, but electric and magnetic fields) as we will describe in later chapters in the book.

The question that Huygens confronted was how to describe and predict the motion of waves. His main interest, of course, was specifically the motion of light waves, but a good starting point as suggested by the epigraph from Huygens that heads this chapter, is the formation of ripples "when one throws a pebble into water and which represent gradual propagation in circles." Huygens made this primitive idea the basis for all wave motion. He described the formation of succeeding wavefronts of a propagating wave, as sketched in Fig. 3, by imagining the wavefront subdivided into very short segments and noting that each segment should produce a circular ripple (we'll call them wavelets), and all these wavelets combine to produce the wavefront along the line that is tangent to all wavelets, producing a wavefront further along the direction of propagation, as sketched in Fig. 3. This procedure of finding the propagation of a wave front is called Huygens' Principle. That's it. It's pretty simple, but very powerful, as we will see when we come back to use the principle many times. (We ignore subtleties such as the issue of backwards propagating wavelets.) Notice that Huygens does not try to explain wavelets in

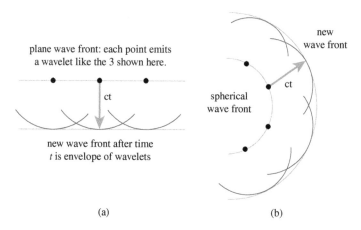

Fig. 3. How Huygens' Principle works. Starting with a wavefront, the wave front at a later time is found by considering each point of the initial wavefront (a few points are shown here) as a source of wavelets that combine to produce the later wave front. (a) A plane (or linear) wave. (b) An expanding spherical (or circular) wave.

terms of forces in the medium. He simply argues that these wavelets should be present in any medium that produces waves.

Huygens found that all he had to do to explain refraction when light enters, say, glass from air was to assume that his light waves travel slower in glass than in air. Then bending takes place because the Huygens wavelets from a wave front move further while they are in air than in glass as illustrated in Fig. 4(a). Thus, when a wave front in air enters glass at an angle, the part of the wave that enters the glass first moves slower than the part that is still in air, causing a cartwheel effect and the bending of the wave direction toward the normal of the surface, again as shown in Fig. 4(a). A marching band uses the same idea to make a turn; the inmost marchers pivot and shorten their steps while the others take progressively longer steps further out, to stay abreast so that each row turns just like a wave front. The basic idea is simple, as many great ideas are.

The speed of light in vacuum, which we will call c, was measured for the first time during the same period that Huygens was producing his theory, as we discuss in Chapter 4, and it was found to be huge, turning out to be about $c = 186,000$ mi/sec. However, as we will see in what follows, the actual value of c is not what mattered for Huygens' theory; instead it's by what factor light travels slower in glass or other material

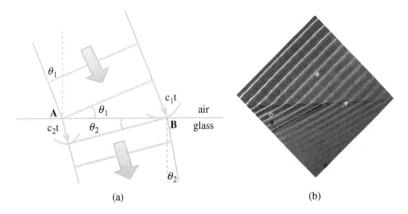

Fig. 4.   Examples of the propagation of a succession of wave fronts changing direction at a boundary where the wave speed c changes. (a) Light waves change direction by Huygens' Principle as they cross from air into glass where the wave speed is slower, $c_2 <$ c_1. (b) Likewise, these photographed water waves change direction as they cross into shallower water where the wave speed is slower.

than in free space, whatever that speed is. Huygens saw that if the speed of light in a material of refractive index n is related to its free space value simply by:

$$c_n = c/n \text{ or equivalently for two materials: } c_1/c_2 = n_2/n_1 \quad (1)$$

then just this one assumption yields the correct bending of light for all incident angles, i.e. yields Snell's Law in Eq. (1) of Chapter 1. We now prove this statement, namely that Huygens' wave theory with Eq. (1) predicts Snell's Law. Referring to Fig. 4, from the definition of the *sine* of an angle in a right triangle (the opposite side divided by the hypotenuse), we see that $\sin \theta_1 = c_1 t / \overline{AB}$ and $\sin \theta_2 = c_2 t / \overline{AB}$, where $\overline{AB}$ is the length of the line joining points A and B. Thus, $\sin \theta_1 / \sin \theta_2 = c_1/c_2$, which Huygens saw yields exactly Snell's Law, $\sin \theta_1 / \sin \theta_2 = n_2/n_1$, if he made just the one assumption that $c_1 /c_2 = n_2/n_1$ as in Eq. (1).[5]

---

[5] At about the same time, in the 1660s, Pierre de Fermat posited his famous *Principle of Least Time* according to which light takes the pathway between two points that requires the least time of travel, at least compared to all nearby paths between the same two points. Fermat showed that light would then obey Snell's Laws and $c_n = c/n$, too. This and similar

Huygens doesn't say why the waves are slowed down in different materials, just that our measurement of the index, i.e. by how much they are bent, also tells us exactly how much they are slowed down; he leaves it to the future to find out why. We will leave it to a much later chapter to find out the answer — namely the end of Chapter 9 when we have learned that light is an electromagnetic wave. But to reiterate, *one great triumph of Huygens' theory could be appreciated back in his time, namely that he predicted the exact form of Snell's Law of Refraction for all incident angles.* Put another way, the first evidence that light is a wave was the form of Snell's Laws. Not conclusive evidence but highly suggestive when Huygens discovered the secret.[6]

Let us mention briefly the then puzzling phenomenon of double refraction which eventually turned out to be pivotal to understanding what kind of wave light actually is. In 1668, the Dane, Rasmus Bartholin, had discovered double refraction in crystals of Iceland Spar, now called calcite. An incident light ray would split into two separate rays in the crystal — one called ordinary, the other extraordinary — and these rays displayed quite different behavior. No one could make any progress in understanding this phenomenon for over a century — except for Huygens. He studied this behavior carefully and once again showed the depth of his genius by finally working out that all observations could be explained by his wave theory if the ordinary and extraordinary beams had different fixed speeds inside the crystal, and if the proportion of the two beams depended upon the angle of incidence relative to a special axis of the crystal known as the optic axis. Huygens could not explain why these two beams existed; that explanation would await Fresnel over a century later, as we will discuss in Chapter 7. Still, Huygens' wave theory of light went much further in accounting for double refraction than any other ideas put forward before Fresnel. We will describe double refraction further and Huygens' ingenious early progress toward explaining it when we take the subject up more fully in Chapter 7.

---

principles are fascinating but require mathematical tools beyond those assumed in this book. Huygens made the point that his wave theory agrees with Fermat's Principle.

[6] Huygens' wave theory predicted Snell's other law as well, that the angle of incidence equals the angle of reflection. The argument is similar to the one above for refraction, but much simpler.

As we will see, light turns out to be a transverse wave somewhat like Fig. 2(a) rather than a longitudinal wave as in Fig. 2(b).

In Paris, and earlier, Huygens had frequent problems of ill-health, sometimes associated with severe bouts of melancholia or depression. During one of these bouts in 1680 he went back to his home in Holland to recover. When he was ready to return to Paris, Louis XIV had revoked the Edict of Nantes in the meantime, and Huygens as a Protestant was not allowed to return. He spent the rest of his life in Holland except for a historic visit to England in 1689, a few years before his death. Huygens had written his brother Constantijn, then in England on a diplomatic mission, that he wished above all to meet Newton. Constantijn arranged the trip; Huygens went to England, and indeed during that visit Huygens and Newton had their one meeting, which we will say more about in the next chapter after we have introduced Newton. Huygens died in 1695 in The Hague.

As we have mentioned already, the wave theory of light fell out of favor for over 100 years after Huygens' time, perhaps due a lot to the commanding reputation of Newton who espoused the rival corpuscular theory. This was despite the elegance and power of Huygens' theory in explaining the precise form of Snell's Law of refraction. One key to eventually resolving the question in favor of waves turned out to be color, but this could only be seen after Newton had placed the study of color on a solid foundation to which we turn next, in Chapter 3. After that we will be able to show how Young and Fresnel in the early 1800s used Huygens' theory as a starting point and Newton's advances on color to show conclusively that light is a wave.

Chapter 3

Newton Turns His Genius
to the Study of Color

"......I could behold
The antechapel where the statue stood
Of Newton with his prism and silent face,
The marble index of a mind for ever
Voyaging through strange seas of Thought, alone."

William Wordsworth's lifelong autobiographical
poem *The Prelude*, published in 1850. These lines refer to
his time at Trinity College, Cambridge.

Color of course is one of the most beautiful features of light and Fig. 1 shows two striking sources of color in nature: (a) a vivid rainbow and (b) iridescence in animals such as butterflies, beetles, and birds. In Chapter 1, we saw how refraction in raindrops explains why the rainbow is always seen at a definite angle near 42°. Now we will see that the appearance of colors in the rainbow is also produced by refraction, in the same raindrops, while in the case of iridescence the colors appear by quite a different mechanism: reflection from very thin films. There are other mechanisms leading to the variety and magnificence of color we see around us, but these two, refraction and thin film reflection, were the subject of much interest and speculation in the 17th century. This speculation led to two remarkable experiments on color, each carried out by Isaac Newton. One was quite famous, in which Newton showed that a white beam of light

(a)   (b)

Fig. 1.   Two natural sources of vivid color studied in the 17th century illustrate two completely different mechanisms that produce color. (a) The rainbow, in which the colors are separated by refraction, as in a prism. (b) A rock dove (the common pigeon) with the colors in its iridescent neck caused by thin films in its plumage. Newton's classic studies on both refraction and thin-film reflection mechanisms are described in the text. Thin-film colors in particular offer an important clue that light is a wave, but the clue was not recognized until the century after Newton.

actually contains all the colors, which are split apart by refraction when the light is passed through a prism (or a raindrop). The other, less famous but even more ingenious, was the production of what are today called Newton's rings that allowed the quantitative study of thin film reflection and, long after Newton, proved to be crucial for the resurrection at last of Huygens' wave theory of light. They both warrant our close attention.

But first let us take a few moments to look at the early life of Newton himself, who often is ranked as one of the two greatest scientists of all time — the other being Albert Einstein. Isaac Newton was born in Woolsthorpe-by-Colsterworth, a hamlet in the county of Lincolnshire on January 4, 1643, three months after the death of his father, a prosperous farmer also named Isaac. At age 12, Newton entered The King's School, Grantham, where he learned Latin but not mathematics. He did distinguish himself at school by building sundials and models of windmills. However, at age 17 he was withdrawn and brought back home by his mother, then a widow for the second time, who wished to make a farmer of him. Newton hated farming and he eventually persuaded his mother to send him back to complete his education. At age 18, he was admitted to

Trinity College, Cambridge, on the recommendation of his maternal uncle the Reverend William Ayscough. Newton paid his way for three years by waiting on tables and taking care of wealthier students' rooms. Then he was awarded a scholarship, which guaranteed him four more years until he would get his MA. After just one year, in 1665, the Great Plague ravaging London reached Cambridge and the University had to close.

Newton returned home for 18 months and devoted himself to study. During that brief time this 22-year-old youth invented the infinitesimal calculus, divined how the motions of the planets and the Moon are controlled by gravity, and started his work on light and color. He had profound insights in all three of these areas and developed them more fully later. Our main interest here of course is light.

At the time, the prevailing theory of color held that color is acquired by a beam of light through interaction with the object the beam enters or reflects from. Descartes, Hooke, and others believed that color is a mixture of light and darkness. In their view, prisms were thought to create colored light and raindrops create the colored rainbow by adding various amounts of darkness as the light passed through. At home in 1666, Newton observed the colored light produced by a crude prism and came to doubt the prevailing theory and instead to suspect that color is an intrinsic property of light so that red light, for example, would stay red light whatever transparent material it is traveling through.

In the years 1670–1672, Newton determined to make a careful study of colored light while also lecturing on optics at Cambridge. With the same skill that later became evident in his telescope lenses, he cut and polished his own prism. This he set up near his window and projected the sunlight passing through it 22 feet onto the far wall, where he observed the full spectrum of colors as in Fig. 2. To prove that the white sunlight actually was composed of all those individual colors, which were then separated from each other by refraction at different angles by the prism (called *dispersion*, with blue light bent the most and red the least), Newton reversed the process. The spreading, multicolored light beam from the prism he sent through a lens to convert it to a converging beam and then into a second prism where the colors were recomposed into white light, as shown *schematically* in one of Newton's sketches in Fig. 2(b). This

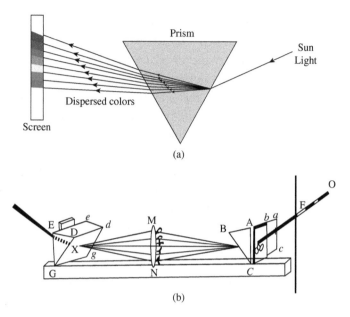

(a)

(b)

Fig. 2. Newton showed (a) that white light is made up of colors and is split into individual colored beams by a prism and (b) that the spreading colored beams so generated by prism ABC are refocused by the lens MN and recombined into white light by a second prism DEG.

recomposing into white light was the crucial experiment and it was not so simple for Newton. Newton's results were a composite of many trials and were very difficult for contemporaries to reproduce. For one thing, they couldn't make such good prisms. Newton nevertheless insisted on the veracity of his results, and backed them up with other experiments to make a compelling case.

For example, Newton went further by separating out a beam that refracted at a definite prismatic angle and shining it on various objects. He showed that regardless of whether a beam was reflected, scattered, or transmitted, it remained the same color. Colors that refract at a definite prismatic angle are often called prismatic colors, and sometimes rainbow colors. All other colors are split by a prism into two or more refracted beams. Some hues produced by two or more prismatic colors look just like another prismatic color; for example, red and green produce a yellow hue that is indistinguishable to our eye from a particular yellow in the rainbow or from the prism. However, Newton also found that other familiar colors produced by two or more prismatic colors are <u>not</u> pure prismatic colors, such as magenta, which is produced by combining the two prismatic colors red and blue but is not itself seen in pure prismatic light. In general, he showed that at least three prismatic colors are required to produce white light. Whether prismatic or not, any color A has a color B that added to A produces white light and is called the *complementary* color of A. Newton's results allowed the specification of any color *objectively* by what prismatic colors form it, the latter specified by what angle they are bent by a standard prism. One hue of red, for instance, can be distinguished from another hue of red by how it is bent by a prism.

It is perhaps difficult for us today to understand how radical Newton's results seemed to his contemporaries, since many of us have heard from childhood that white light is composed of all the colors and a prism simply splits them apart. But at the time, this was a startlingly new idea to most people, including great scientists like Descartes and Hooke who held that color was *created* by the interaction of light with substances, as pointed out above. Of course, Newton became famous among astronomers and other natural philosophers for his general laws of motion, his Law of Gravitation, and his explanation for the orbits of the planets. But among the rest of the educated public, his experiments with the prism were a major part of his fame for centuries, as illustrated by the statue at Cambridge described by Wordsworth in the epigraph heading this chapter.

Newton argued that the colors of the rainbow are similarly produced from white sunlight just as colors had been revealed in his laboratory by his prism, but in the case of the rainbow, by refraction in water drops in the atmosphere. Light entering water shows dispersion as it does with glass, so that colors with a greater index, e.g., blue, bent more and those with a lesser index, e.g., red, bent less. Thus, sunlight entering a water drop gets split into different colored rays; using the terminology of Chapter 1, the Descartes Ray follows different trajectories for different colors and thus the rainbow occurs at a slightly different angle for each color. Using his measured dispersion of water, Newton calculated correctly the spread of colors actually observed in the rainbow.

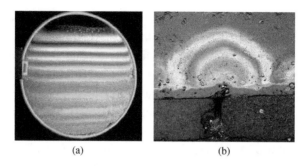

(a)   (b)

Fig. 3. Two sources of brilliant colors caused by reflection of white light from thin films, producing somewhat similar color patterns. Such sources, as well as the plumage of birds as in Fig. 1, were common in the 17th century, and were studied by Hooke, Newton, and others. (a) A soap film; the film gradually gets thicker from top to bottom. (b) A film of oil on water covering a stone or sidewalk, the thickness of the oil getting thinner as it spreads outward from the source at the bottom. Clearly, in (a) and (b) a change in color is associated with a change in film thickness, but the precise thickness is difficult to measure. Newton solved this problem with *Newton's Rings*.

We will change the subject now and turn to the even more remarkable experiments mentioned already that yielded Newton's rings. They began with the fascinating and vivid colors that are seen from thin films. Examples include bird plumage, as was seen in Fig. 1, and soap films and oil films as shown in Fig. 3. It was well known by Newton's time that when white light encounters such films, the reflected light is colored. The actual color depends strongly on the thickness of the film as explained in the caption of Fig. 3. Figure 4 shows a 1674 painting of a girl having blown a soap bubble, a thin film which the artist depicts with touches of color doubtless familiar to his viewers.

If a drop of oil is deposited on water, and starts spreading out over the surface, the color is observed to change quickly as the oil film gets thinner and thinner, usually ending up bluish just before the film becomes too thin to be seen. Robert Hooke studied this effect by placing one glass plate on top of another to create a thin air film between them. The air film could be made thinner or thicker by pushing down on the upper plate with a greater or lesser force. Hooke published a discussion of this work in his treatise *Micrographica*, which Newton studied carefully in 1670. Newton became convinced the color was *not*

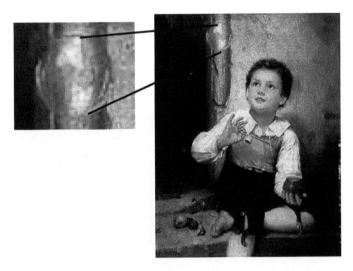

Fig. 4. *Girl Blowing Soap Bubbles* painted in 1674 by Pierre Mignard, showing that soap films and their colors were familiar sights in the 17th century. The artist not only captures the fascination on the girl's face, but also takes the trouble to show part of the reason she is fascinated, the thin-film colors on the left side of the bubble where there is no competition from background light. These colors are similar to those of the soap film in Fig. 3(a).

produced by the prism effect already discussed, and set out to study it carefully. Newton saw that Hooke was unable to determine the dependence of the reflected color on the thickness of the air film, because he could not measure the film thickness.

Newton then had an ingenious idea for measuring the film thickness. Rather than using two flat surfaces, one could use one flat surface and the curved surface of a lens touching the flat surface to create the air space, as shown in Fig. 5. The thickness of the air film would be zero at the point of contact and increase with distance away from that point by an amount that could be calculated from the measured curvature of the lens. Newton not only had the insight to *conceive* this experiment, but he had the ingenuity and determination to actually *do* it. He took one of the excellent telescope lenses he had made already, placed its convex surface on a flat surface of equally good glass and pressed them together with an adjustable clamp. He then studied the reflections when the air film was illuminated from its flat end.

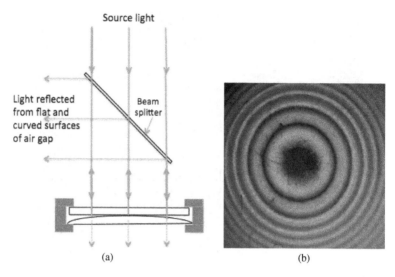

(a)   (b)

Fig. 5. Newton's Rings produced by a glass lens and flat plate clamped together as shown in (a). A thin film of air (greatly exaggerated in the illustration) is formed between the two glass surfaces. The ensemble is illuminated from above, producing reflected light that is viewed as shown. The reflections of interest are those from the surfaces forming the air film. When the source is white light, the reflected light contains colored rings as shown in (b). When the source is a single color, rings of that color appear separated by dark rings of no light.

When he illuminated with white light, the reflected light formed a series of colored concentric rings centered at the point of contact between the surfaces, as displayed in Fig. 5. He wrote down the color of each ring in the reflected light and the thickness of the corresponding air space. He did the same for the transmitted light. The colors of the rings of transmitted light were complementary to the corresponding reflected colors, indicating that together they account for the total incident white light. Using just light of one color, he could distinguish about 30 successive rings, as will be illustrated for yellow light in Chapter 6 where we present Thomas Young's explanation of the rings by the wave theory of light over 100 years later. Newton found that the diameter of violet rings was smaller than that of the corresponding rings of red light, meaning violet light occurred with a smaller air film thickness than red light. Green light was intermediate between violet and red.

As Newton stated in his publication *Opticks*:

"Obs. 13. Appointing an Assistant to move the Prism to and fro about its Axis, that all the Colours might successively fall on that part of the Paper which I saw by Reflexion from that part of the Glasses, where the Circles appear'd, so that all the Colours might be successively reflected from the Circles to my Eye, whilst I held it immovable, I found the Circles which the red Light made to be manifestly bigger than those which were made by the blue and violet. And it was very pleasant to see them gradually swell or contract accordingly as the Colour of the Light was changed. The Interval of the Glasses at any of the Rings when they were made by the utmost red Light, was to their Interval at the same Ring when made by the utmost violet, greater than as 3 to 2, and less than as 13 to 8. By the most of my Observations it was as 14 to 9. And this Proportion seem'd very nearly the same in all Obliquities of my Eye ..."

Clearly, each color had its own air film thicknesses. A striking feature was the regularity of the thicknesses for successive rings; the thickness increased the same amount between one colored ring and the next of the same color. For successive violet rings this increase in air gap thickness was 210 nm (1 nm = 1 nanometer = 1 billionth of a meter) and for successive red rings it was 330 nm, with the other colors falling between in the same order as seen in the prism or a rainbow. These results are not only intriguing but the precision of measurement is spectacular, and is a tribute to the amazing experimental ability of Newton. These original results provided the most accurate measurement of the wavelength of light when at last Newton's Rings were correctly interpreted. His experiments on light stood alone in ingenuity and precision for over a century and a half. Clearly, in addition to all his analytical gifts and conceptual insights, Newton possessed great experimental skill. In lens grinding, for instance, the shape of the curved surface had to be accurate to better than a tenth of a micron to make the distinct pattern of rings and to enable the measurements of the thickness of the air film at each ring diameter.

Newton was fascinated by the regularities in his results, and toyed with musical analogies, especially frequency intervals between musical tones, though he did not follow up on any musical ideas. However, Newton did propose a possible explanation for the rings, based on what he called "fits", i.e. waves similar to sound waves created in a hypothetical medium by the light crossing the glass/air surface, much as a stone entering water creates ripples. The fits would increase or decrease the transmission and reflection of light at the next interface, depending upon how far they had traveled in the air film, hence the dependence on the air film thickness. Actually, the fits were part of Newton's complete theory of refraction, and they had the perceived advantage over Huygens' wave theory that they did not require a medium permeating all space to carry the waves, but only in the refracting material itself to carry the "fits". Although fits were worked on by many devotees of Newton for the next 100 years, they did not lead to significant advances and we will not discuss fits further.

So Newton's rings remained a mystery for over a hundred years until finally Thomas Young came up with the answer using the wave theory of light. Thus, the greatest proponent of the particle theory left the world a deep mystery that eventually could only be solved by Huygens' wave theory.

Fig. 6.   Portrait of Isaac Newton at 46 by Godfrey Kneller.

Newton's Later Years

Prompted by correspondence with Hooke in the early 1680s, Newton revisited his investigations of gravity and motion of the 1660s, and then at the urging of Edmund Halley began to write them down in a systematic fashion. In this way Newton's work on motion and forces, begun during the plague years he spent at his mother's home, evolved into the creation of the *Principia*, and its publication in Latin in 1687, one of the most celebrated works in history. It covered all known physics except for his own discoveries in optics, which he published separately, and included his three basic laws governing all bodies in motion[1] and the inverse square law of universal gravitation. It explained the orbits of the planets about the Sun, the Moon about the Earth, Jupiter's moons about it, and in principle how to calculate any motion on Earth or in the heavens once one knew the forces.

Newton was often involved in controversies about priority. As soon as the *Principia* was published, Robert Hooke accused Newton of stealing the inverse square theory, claiming that he, Hooke, had conceived the theory years earlier. In fact, Hooke had mentioned the idea, but had never done anything to prove it, and most scientists ignored Hooke's claims. Not Newton. He was outraged, publicly defended his discoveries as his own work, and withdrew all mention of Hooke from his notes and future publications of the *Principia* and other works. At the behest of Halley, Newton eventually agreed to insert an acknowledgment of Hooke's work (together with that of Wren and Halley) in the discussion of inverse squares, but that did not at all placate Hooke. So the feud continued until Hooke's death in 1703. In fact, Newton did not publish his book *Opticks*, containing his work on color we have been discussing here, until Hooke's death, apparently to avoid further controversy.

Another famous controversy was with Leibniz over the invention of the infinitesimal calculus. Leibniz publicly claimed priority and accused Newton of taking his ideas and not acknowledging the debt to Leibniz. In fact, Newton had developed and used the equivalent ideas, what he called Fluxions, back in the 1660s during the plague years mentioned already, which probably gave him priority, but he never published them until the

(Continued)

---

[1] (1) A body maintains its state of motion or rest unless acted upon by an external force; (2) Any change in motion is proportional to the external force applied; and (3) For every action, there is an equal and opposite reaction.

(Continued)

Principia. It is now clear that the two men developed these ideas completely independently. So Newton engaged in very public quarrels, but in fact he disliked them, and tried to avoid them, as in not publishing his *Opticks* until Hooke's death. So indeed he rarely started controversies, but once aroused he was bitter and thorough in counterattack, and never called a halt to it.

It is noteworthy that Newton and Huygens never engaged in a long controversy, and in fact Newton went out of his way to acknowledge Huygens' priority in solving the circular motion problem in connection with the pendulum clock as mentioned in Chapter 2, and Huygens quickly admitted his error when he had misconstrued some of Newton's results on prism refraction.

In addition to his positions in the Royal Society (as its head from 1704), Newton held public office many times. He served for two years as a Member of Parliament for Cambridge University. His most famous public service was at the Royal Mint, first as Warden starting in 1896 and then as Master of the Mint in 1699, arranged as a sinecure by Charles Montagu, 1st Earl of Halifax, then Chancellor of the Exchequer. Newton, however, took these positions very seriously and remained at the Mint for the last 30 years of his life. He especially cracked down on counterfeiters, even frequenting bars under disguise to gather evidence, conducted over 100 cross-examinations of witnesses and suspects, and successfully prosecuted 28 coiners — who were then executed for high treason. Newton was also instrumental in reforming the currency and switching from a silver to a gold standard.

Let us now go back to just one more incident in Newton's life: when he and Huygens met for the only time, on Huygens' last visit to England in 1689. Huygens just prior to the visit had expressed a strong desire to meet Newton, whom he admired greatly. In fact, Huygens had received a copy of the *Principia* less than a year earlier and had been spending long hours studying it, especially the parts on gravitation.

Huygens' brother Constantijn was in England already, having come over from Holland as a high diplomat with William and Mary in 1688 when they became king and queen at the invitation of Parliament in the "Glorious Revolution". Constantijn's journal is a source of information about military and political events, of William's crossing and accession to the throne as well as a collection of everyday — often humorous and earthy — observations and stories. Constantijn helped arrange Christiaan's visit in early June 1689. Newton and Huygens attended a short meeting of the Royal Society

(Continued)

in London in June. The irony is that Newton spoke on double refraction not having yet read Huygens' much deeper treatment, while Huygens spoke on gravitation without having finished the *Principia* in which Newton shows much deeper analysis. We do not know if they spoke with each other at the meeting, but we do know that they met for a carriage journey together in July. Constantijn records in his journal that Christiaan left for London at 7 AM on July 10, 1689, with Isaac Newton together with Nicolas Fatio, a close scientific acquaintance of Huygens', and John Hamden, who was well known to Newton. Fatio, who had an especially good command of French and English, together with Hamden could serve as interpreters to facilitate the conversation between Huygens and Newton. Constantijn does not say where they boarded the carriage, but it was certainly at least a half day from their destination in central London, probably the Royal Society at Gresham College. In any event, Newton and Huygens must have had extensive conversations during this journey. In fact, Huygens later wrote to Leibniz that Newton had told him of remarkable experiments on color, almost certainly Newton's rings, because Huygens already knew about the prism results.

Imagine anyone observing this carriage speeding along. They could hardly guess there were two of the most powerful intellects of all time conversing for the first time inside. Huygens moved in elite European society from birth; Newton was the son of a farmer and had to wait on other students during his first three years at Cambridge. A skillful dramatist could perhaps write a one-act play to convey the human, emotional, and intellectual power or rivalry in the dialogue during this historic journey. We will leave all this to the reader's imagination.

So, we leave Newton and Huygens. What about their respective particle and wave theories of light? Huygens' wave theory accounted for reflection and refraction in a more complete and quantitative way. It explained Snell's Law for all incident angles with just the assumption that the speed of light in a medium of refractive index n is c/n. However, the wave theory seemed to present difficulties, which were stated most clearly by Newton: (1) It seemed to require a medium for the waves, the so-called aether, which permeated all bodies and all empty space and carried light waves but was so subtle it did not hinder the motion of even the lightest bodies. (2) Light does not seem to bend around obstacles the way sound waves and water waves do, but instead casts sharp shadows. We will see how these difficulties were finally answered in later chapters.

Whatever their relative merits, the particle theory reigned supreme for over 100 years, certainly in large part because of the reverence for Newton. Then the fortunes began to reverse, as we will see. But in the meantime, astronomers were beginning to make observations that revealed for the first time how fast light travels. Whether it was particles or waves, its speed was unbelievably fast! That is the story we tell in the next two chapters before returning for the triumph of the wave theory.

Chapter 4

The Speed of Light 1: Timing Jupiter's Moons. Paris 1671–1676

"Philosophers have been laboring for many years to decide by some Experience, whether the action of Light be conveyed in an instance to distant places, or whether it requireth time."

Ole Rømer, *Journal de Scrivans* (Paris 1676).

On November 21, 1676, Ole Rømer, a young Danish astronomer working at the Paris Royal Observatory, stood before the French Royal Academy of Sciences and announced that Observatory astronomers had measured the speed of light. He said that observations of Jupiter's moon Io showed that light travels through space at an incredibly fast speed, but a speed the Observatory astronomers could actually measure. At that moment, no one outside the Academy had the slightest idea how fast light travels; some even thought its propagation between two places to be instantaneous. And no one would have dreamed that the speed of light could be found, as Rømer told the Academy, by carefully timing the regular eclipses of Io by Jupiter and noting apparent delays when the Earth moved further away from Jupiter and Io, increasing the length of the light path to Earth. This discovery was made by a truly international team of astronomers — the first ever assembled at Paris or anywhere else. In fact, it is the first example known to the author of what is called today "Big Science", and we will devote some time to it.

To add to the drama at the Academy meeting, there was immediate and vigorous objection to Rømer conclusions by no less than the Director of the Observatory and a favorite of Louis XIV at Versailles, the distinguished Italian astronomer Giovanni Domenico Cassini. Thus began a long dispute over what eventually came to be recognized as a true measurement of the speed of light, commemorated by this inscription on the north frontage of the Paris Observatory: "The Danish astronomer Olaus Rømer (1644–1710) discovered the velocity of propagation of light at the Paris Observatory in 1676." Since it is now understood that the speed of light should be the same for all observers anywhere, the inscription at Paris in fact refers to the first measurement of a universal quantity made on Earth.[1]

Rømer explained in his talk to the Academy, and more fully in a paper published the following month, how Io eclipses appear later as the Earth in its orbit about the Sun retreats from Jupiter and Io, the apparent delay being due to the time required for the light from Io to travel the extra distance to the Earth. Rømer and his audience knew that, if true, this discovery was a major advance in knowledge. And yet they could have no idea how important the speed of light would come to be. Nor for that matter, how familiar it would become to the public. In addition to visible light, many other forms of electromagnetic radiation are well-known — from radio waves to gamma rays — that all travel with this same speed in free space. The *light year* is familiar as a unit to measure distances to the stars. At a fundamental level, many people have at least heard of the cardinal tenets of Eintein's Theory of Relativity that no object or signal can travel faster than light, and that light has the same speed relative to all observers regardless of their speeds relative to each other. And there's the famous equation $E = mc^2$ giving the conversion of mass into energy, where c is the speed of light, telling us that this speed plays a crucial role in the basic laws of physics.

---

[1] One intriguing way of seeing the significance of c being a universal quantity is to suppose (as in SETI, the *Search for Extraterrestrial Intelligence*) we broadcast a message for some distant intelligent life to detect and decipher, and we wish to tell them how fast our jet planes travel. They would not know our units of distance or time, so saying 550 miles/hour would be meaningless. But if we said the speed is one millionth the speed of light, they could figure it out readily in their units since c is a universal quantity.

To Rømer and colleagues, what was most striking was simply how unimaginably fast light travels, taking only tiny fractions of a second to traverse ordinary distances; the second sentence of Rømer's paper informs us that even for 8,000 miles, about the size of the Earth, "light needs not one second of time." Rømer's discovery was possible because of the vastness of the solar system; light, even with its great speed, needs a measurable interval of time — several minutes — to traverse distances comparable to the 93-million-mile radius of the Earth's orbit about the Sun. We now know that light from the Sun takes about eight minutes to travel this radius and get to the Earth; Rømer inferred a time of 11 minutes from the Paris measurements, which corresponds to a speed of 140,000 miles per second, reasonably close to the correct speed of 186,000 miles per second. Yet, even this modest discrepancy, long thought to be due to measurement errors in Paris or errors in Rømer's analysis, we now know reveals a remarkable effect of Jupiter's moons lurking in the Paris data that is likely to surprise even some of our more learned readers when we reach the point of discussing the observations in detail. Both the Paris observations and Rømer's analysis of them turn out to be exquisite even by modern standards, and close study of the 1670s data will tell us how Jupiter's moons actually kept time *with each other* then, as they still do today with intriguing effects on their geology and landscape (and even potential for life) currently under extensive study.

There was an additional touch of drama in Paris. As Rømer pointed out in his paper the astronomers had made a daring prediction starting in August, three months prior to the November 21 announcement: they claimed publicly that the eclipses of Io in the coming November would take place about 10 minutes later than the time one would think based on eclipses observed in August when the Earth was much closer to Jupiter and Io.

This prediction had followed several years of observations starting in 1671 during which they saw small perplexing deviations in the Io period. It was almost certainly young Rømer who had then conceived the idea that the deviations were due to *mora lumis*, the finite speed of light, for Cassini himself said years afterwards, in 1693: "Monsieur Rømer explained very ingeniously one of these inequalities that he observed for several years in the first satellite [i.e. the innermost moon Io] by the successive motion of light which needs more time to come from Jupiter to the Earth when it is more distant than when it is closer ..." (Cassini then goes on to explain why he doubted this "ingenious" explanation, as we will discuss in due course).[2]

As it turned out, a special observation was indeed made on the evening of November 9, when Jean Picard, the "father of French astronomy", together with his former protégé Rømer and others from the Royal Observatory made measurements to test the very public prediction. We will come back and look closely at this night as a good example of how astronomers actually used telescopes to measure the eclipses so accurately.

What made such exquisite results possible? Certainly the telescopes did, in this brief era of super-long ones, some stretching to hundreds of feet, and the skill with which they were used. Crucial also were the bold and brilliant insight and analysis. Rømer's result was accepted in many places, particularly in England (where it was embraced by Edmund Halley known today for the comet that bears his name), but not in Paris for over 40 years. Finally, confirmation came from quite a different type of observation by the British astronomer James Bradley in 1729, the story we will tell in Chapter 5.

Over 60 years earlier, in 1609, Galileo had turned his telescope to the heavens and immediately reported startling observations such as dark spots on the Sun and mountains on the Moon. Then, in January 1610 came his most famous discovery, the moons of Jupiter. He saw the four largest, known as the Galilean moons, shown in Fig. 1. He found that they orbit Jupiter with quite regular periods, that of the innermost moon Io being

---

[2]Author's note: Despite Cassini's statement crediting Rømer with the discovery, there is a minority opinion that Cassini himself originally conceived the idea, based mainly on a memo written by Cassini as Director, alerting the Academy to the eclipse delays and giving the finite speed of light as one of the possible explanations, though not attributing the idea to anyone. The memo is now known to have been issued in August 1676, just three months before Cassini's adamant public opposition to the speed of light idea and Rømer's presentation of detailed arguments and calculations that as a practical matter he must have begun long before August 1676. It thus seems most reasonable to accept Cassini at his word in 1693 that the originator of the idea was Rømer. As everyone agrees, Rømer is the one who "very ingeniously" carried out the published analysis backing up the idea.

Fig. 1.   Jupiter and the four Galilean moons seen from Earth. Io is the closest one to Jupiter.

close to 1.7 days. The moons are seen by reflected sunlight, and Galileo noticed that as they orbit Jupiter, they undergo regular eclipses when they pass through Jupiter's shadow once each orbital period — just as our Moon is eclipsed, although relatively rarely, by the Earth's shadow. Jupiter is so large and casts such a large shadow that its moons never fail to pass through its shadow and undergo an eclipse. The situation is sketched for Io in Fig. 2. As we watch Io from Earth, once every 1.7 days it disappears for about two and a quarter hours in Jupiter's shadow.

So why were astronomers in the 1660s and 1670s measuring these eclipses of Jupiter's moons so carefully? They of course had no notion about the speed of light at first. They were working on an important practical problem, how to determine the longitude (i.e. the east–west position) of any location on Earth. For ships at sea, knowing their longitude was of course crucial. On land too, longitude was necessary for accurate surveys of area. What was needed for longitude was an accurate clock at any location of interest to compare with the positions of the Sun or stars. Indeed, airplane passengers are used to turning back their watches if they fly west. If you set your watch at noon in Paris (which is when the Sun is crossing what astronomers call the local meridian and reaching its highest point there) and then fly to New York, you will notice the next day that your

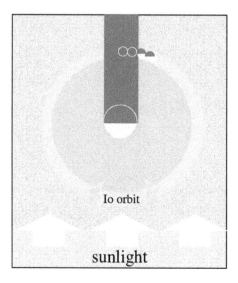

Fig. 2. Emersion: Io emerging from Jupiter's shadow. From Earth this orbit is seen edge-on.

watch on Paris time reads close to 6 PM when it is just noon in New York and the Sun is just then reaching its highest point. You could then know that you have moved a quarter the way around the Earth and your longitude in New York is 90° west of Paris (since the Earth turns 360° in 24 hours = 90° in six hours). At night, you could use a star crossing the meridian as a reference instead of the Sun. Notice you need your *watch*; and in the 17th century man-made timepieces were not accurate enough, and especially not for the long duration of travel then needed (for example, by stagecoach and ship from Paris to New York). But as Galileo, the discoverer of Jupiter's moons, had pointed out, the regular periods of those moons provided an accurate clock that could be seen from anywhere on Earth and used to synchronize to a common time everywhere.

The astronomers concentrated on Io which has a shorter orbital period than the other moons; the greater frequency of eclipses made it the most valuable for the longitude problem. Io's period could be measured most accurately by timing when the moon either entered Jupiter's shadow (an immersion) or emerged from the shadow (an emersion). An emersion of Io is sketched in Fig. 2. An emersion or immersion of Io takes about three minutes, and by measuring the time between successive immersions or successive emersions it was thought that a skillful observer might be able to use the best telescopes of the 1660s and 1670s to obtain the 1.7 day period of Io to an accuracy of about one minute. (In fact, we now know the Paris astronomers did a little better than this.) Since Picard already had measured the size of the Earth accurately using the change in star elevations with distance measured along a north–south line, it was known that at mid-latitudes the Earth's rotation causes the surface to move about 12 miles in one minute, which indicates an uncertainty in east–west distance as small as 12 miles could be achieved with Io. This accuracy was much better than any other method of the day. Even with the irregularities already mentioned that had been seen in the eclipse times from one month to the next, errors in longitude could be avoided by comparing eclipse times taken within one or two weeks of each other at the two locations. (One could use a less capable telescope elsewhere to get the Longitude relative to Paris by picking a time when the eclipses were well-resolved from Jupiter's disk.)[3]

The Paris astronomers used their work on Longitude and Latitude during this period to measure the size of France accurately for the first time. Cassini sent a team with telescopes to the major cities of France where they observed eclipses of Io and recorded the location of a reference star at the time of eclipse. Comparing notes later, if the star had reached this location 20 minutes before the same eclipse in Paris, they had been about $12 \times 20 = 240$ miles (Longitude close to 5°) west of Paris. The results were a big surprise. Coastal cities in Normandy, for example, thought to be over 400 miles west of Paris, reached by roads trod on for over a thousand years dating back to the Romans, were closer by 60 miles or more than the old maps had indicated. The country turned out to be significantly smaller than expected, which made a big impression on Louis XIV who, as we will see, took a keen interest in the Royal Observatory. On one occasion he is quoted as saying; "I pay my astronomers well and they have *diminished* my kingdom." On another he quipped: "Cassini has taken more of my kingdom from me than I have won in all my wars."

---

[3] See Fig. 8 later in this chapter.

The work on Longitude had led directly to Rømer's coming to Paris. At the time perhaps the most detailed and precise star maps had been compiled at the observatory castle Uraniborg near Copenhagen, starting with the great Tycho Brahe who had founded Uraniborg in 1581 (and pinpointed the planets so precisely that Kepler could show their orbits are ellipses with specific properties, which in turn Newton explained with his Law of Universal Gravitation). To make full use of these tables at the Paris Observatory, they required a complete copy of them, and also needed to know the relative Longitude of the two locations. To this end, Jean Picard traveled to Denmark to observe Jupiter's moons near Uraniborg and compare with Cassini's observations at Paris at the same time. After a while Picard wrote that a remarkable young Danish assistant named Rømer was of immense help in making the observations, and also had figured out on his own how to update Cassini's calculations as to the approximate time of suitable eclipses in the immediate future. Picard strongly urged that the brilliant Rømer be hired at Paris. Cassini applied for a position and the funds were granted by the king, so Rømer came to Paris with Picard in 1671. In his notes, Picard wrote that he brought back the eight month journal of eclipses plus *two* treasures, the book of Tycho's Uraniborg observations and Rømer. So, to the strong team of astronomers in Paris, already the most eminent in Europe, came a little known young man from the north who quickly established himself as a powerhouse and in the space of seven years left an enduring mark there and on the history of science.

Ole Christensen Rømer was born in Aarhus, Denmark, on September 25, 1644, into a merchant family of modest means. He must have distinguished himself in some way, because he matriculated in 1662 at the University of Copenhagen, at which his mentor was Rasmus Bartholin an eminent scientist who published his discovery of the double refraction of light by Iceland spar (calcite) in 1668 while Rømer was living in his home. (We mentioned this discovery by Bartholin in Chapter 2 where we saw that Huygens made the first progress toward understanding the phenomenon with his wave theory of light.) Rømer was given every opportunity to learn mathematics and astronomy using Tycho Brahe's astronomical observations, as Bartholin had undertaken the task of preparing them for publication, which would yield the book later provided to Picard. It was also of considerable importance to Rømer, as it turned out, that while living in Bartholin's house, he came to know Bartholin's daughter, Anne Marie Bartholin. When Rømer returned to Denmark after 7 years in Paris, he and Anne Marie were married, and Rømer lived the rest of his life in Denmark, achieving fame in both science and public life.

By contrast with the relatively unknown Rømer, Cassini was already an internationally renowned astronomer when he arrived in Paris in 1669. Born on June 8, 1625, in the Republic of Genova, Cassini became a devotee of astrology as a young man. In 1645, he was hired by a senator of Bologna, the Marquis Cornelio Malvasia, to work at the Panzano Observatory then under construction. He and Malvasia spent much of their time improving the ephemerides for astrology, but Cassini's interests were shifting to astronomy, and he was studying Jupiter's moons already at that time. In 1650, Cassini became Professor of Astronomy at the University of Bologna. He joined the telescope maker Campani to make many planetary observations, measuring rotation periods of Mars and Jupiter. The connection with Campani would prove a fruitful one for many years to come, as we shall see. Meanwhile, in 1669, Cassini was offered the job as director of the new Paris Observatory, which he helped set up and opened in 1671. Officially he served a dual role, as astronomer and astrologer to Louis XIV ("The Sun King"), though he focused the majority of his time on astronomy.

Well before this time Jean Picard, the other astronomer so crucial to our story, had improved the practice of observations in Paris to the point that he is now regarded as the founder of modern astronomy in France. He was an early advocate of using astronomy in surveying, and his careful measurement of the Earth's diameter, accurate to a fraction of a percent, was used not only to convert latitude to distance but also by Newton to validate the inverse square law of gravity.

Picard was born on July 21, 1620, to a bookseller in La Flèche, a town on the banks of the Loire. Jean attended the Jesuit College at La Flèche, founded under Henry IV in 1603 with the instructions "to select and train the best minds of the time." A famous graduate "of the time" was René Descartes whom we have already met in Chapter 1 (and who, ironically enough, advocated the view that the propagation of light is "instantaneous") and the College quickly gained a reputation as one of the best educational establishments in France. Around 1644 Picard went to live in Paris, where he became an assistant to Pierre Gassendi, astronomer and professor of mathematics at the Collège Royale, and his merit in observational astronomy became well recognized. Picard referred to himself as the Abbé of Rillé, a town near La Flèche. He may simply have been an abbé commendataire i.e. a layman who drew the revenues of a benefice that provided an income. The general view is that Jean Picard was a shy and modest abbot, a tireless worker, often away in the provinces or abroad on some important project while others were in the limelight in Paris. Nevertheless, in 1655 Picard became professor of astronomy at the Collège de France in Paris.

Picard was made a founding member of the French Royal Academy of Sciences when it was established in 1666 by Louis XIV at the suggestion of his influential Finance Minister, Jean-Baptiste Colbert. Three years later Picard, considered one of the greatest observational astronomers of the day, would have been a natural candidate for the directorship of the new Royal Observatory, but the story goes that he was modest and unselfish enough to recommend the rival Italian astronomer Cassini to Colbert and Louis XIV. Picard may have recognized that he did not possess Cassini's dazzling presence which was to prove so effective at the court. In the new observatory Picard remained a tireless worker whose passion was precision, in which he was probably unsurpassed in his generation throughout Europe. His focus on precision and his ability to achieve it proved indispensable in obtaining the exquisite observations that were the basis for Rømer's discovery. These observations were likely a team effort, but most of them are recorded with Picard's signature and bear the stamp of his approval, skill, and dedication.

When Cassini came to Paris to become Observatory Director, he brought the 17-foot telescope given to him by Campani, his collaborator in Italy and widely considered the best telescope maker of the day. The excellence of this instrument impressed the astronomers of the French Academy, and also Colbert. In consequence, a 34-foot telescope was ordered from Campani and presented by Louis XIV to the newly erected observatory of Paris. Figure 4 shows a painting of Louis XIV receiving thanks from Colbert for founding the Academy and the Observatory. Many academy members are shown (note Cassini and Huygens in full

Fig. 3.   Cassini, Huygens, Picard.

Fig. 4.   This painting by Henri Testelin, Secretary of the Royal Academy of Painting and Sculpture, shows Louis XIV receiving thanks from Colbert on behalf of the French Academy and its members. The 17-foot telescope is shown in the window behind Colbert's head and the 34-foot is on the floor at the very front. The date is before Rømer joined the Academy. Cassini, Huygens, and Picard (Picard attired quite differently!) are shown together, and in the dimness behind them is a fossil ape-skeleton. The painting now hangs at Versailles.

regalia, Picard in the plain attire befitting an abbot) as are both the 17-foot and 34-foot telescopes, the former through the window hanging on the observatory grounds behind Colbert's head, the latter on the floor (!) in front of the King, decorated with fleur de lys figures. The 34-foot telescope was used for all the observations of Jupiter's moon Io that were the basis of the measurement of the speed of light.

The French Academy of Sciences was indeed a remarkable learned society which right from the beginning had foreign members. The Dutchman Christiaan Huygens, the founder of the wave theory of light and next to Newton perhaps the greatest physical scientist of the day, was one of the founding members. He was granted an apartment in the Academy building and the largest stipend in the Academy until Cassini arrived and received double Huygens's stipend. The Academy not only encouraged scientific research among the French, but also attracted brilliant scientists such as Huygens to France who contributed to pure science and to practical advances. Rømer affords an excellent illustration. He was made a member of the Academy in 1672 and lived in the Observatory (where he shared Picard's rooms). There, in addition to his work on Longitude and the speed of light, he invented a micrometer that was adopted for general use, developed hydraulics instruments, and took part in the levelings for conduction of water to Versailles and elsewhere. Rømer also demonstrated in a lecture to the Academy that the best shape for the teeth of gears is the epicycloid, which was then quickly applied both for hydraulic machines to lift great weights and for the intricate workings of Huygens's superb pendulum clocks, one of which was made for precise timing at the Observatory. Besides all this, as we described in Chapter 2, Rømer worked with Huygens on an efficient design of the microscope that was universally adopted. As Colbert had foreseen, the Academy would more than repay the great investment of livres with valuable knowledge and applications of science.

Now let us turn to the telescopes! Such great ones as shown in the painting made the discovery about the speed of light possible. Key to a good telescope is the primary lens at the front, which gathers light from a distant object such as a tree and focuses it to form an image that can be viewed up close and enlarged with a magnifier (the eyepiece) as shown in Fig. 5. The image will be larger the longer the focal length F_1 of the primary lens and will be magnified when seen though the eyepiece, the more so the

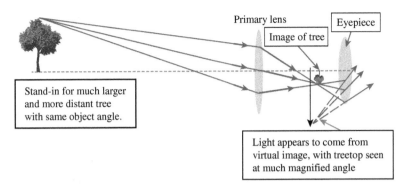

Fig. 5. Refracting telescope sketch. The primary lens creates an image of an object which is magnified by looking at it with a very close eyepiece.

closer it can be to the eyepiece, i.e. the shorter the eyepiece focal length F_2. The total angular magnification is thus $M = F_1/F_2$. The larger the primary lens, the more light is intercepted and the brighter the image. In early telescopes, imperfections in the primary would blur the images, making them overlap and limiting the gains that could be made with longer focal lengths. Also, such imperfections limited the usable area of the primary, cutting down on the image brightness and the visibility of faint objects.

Telescope lenses became much better between 1610 and 1660, as demonstrated in recent times by comparative tests of lenses in famous telescopes of that period. Lens grinding and polishing improved, as we have already seen in the excellent lenses fabricated by Huygens and Newton. The best of the lenses tested in modern times were fabricated by the Italian telescope maker, Giuseppe Campani, who as we have learned supplied the great telescopes at the Paris Observatory. He was certainly a skillful lens grinder, but the superiority of his lenses was at least partly due to the quality of glass available in Italy.[4]

---

[4] In fact, the Italians were famous for their expertise in glass work. At the same time as Louis XIV funded the Royal Observatory, he was getting ready for another, much larger project with glass at Versailles, what became his famous Hall of Mirrors, and again the Italian artisans were the best. In the 17th century, the Venetian Republic held the monopoly on the manufacture of mirrors. Colbert, to fulfill his mercantilist principles, enticed several workers from Venice to make mirrors in France. Apparently, the government of the Venetian Republic sent agents to France to try to poison the imported workers; nonetheless, the mirrors were in fact fabricated in France.

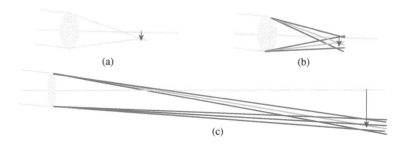

Fig. 6.   Chromatic Aberration (CA) (a) the image of a distant object by a perfect lens; (b) the effect of dispersion of the lens in smearing the image at the focus. This is CA; (c) effect of narrowing the lens to increase the focal length where the spread in colors at the focus is the same as in (b) but the image size is larger and resolved because of the increased focal length. Hence, the super-long telescopes.

Now, what about those ultra-long telescopes? Once lenses became good enough, the most serious problem that limited telescopes in the mid-1600s was "chromatic aberration", the problem that a simple telescope lens — like a prism — spreads white light into a rainbow of colors and blurs the image. We have already seen the dispersion of a prism in Fig. 2 of the previous chapter. Figure 6 shows the similar dispersion of a lens; blue light is bent the most, red light the least, and greenish yellow somewhere in the middle just as in a prism. The figure shows how the dispersion smears out the image at the focus of a lens; at the point where green/yellow has a sharp focus, blue and red will be spread out over a circle about 1/50th the diameter of the lens. This fraction is independent of the focal length (i.e. a thinner lens spreads the colors by a smaller angle but through a longer focal length since the lens is thinner, yielding the same final spread). The center will appear green/yellow and the rim magenta. Telescopes do better than this because, as Newton pointed out, the eye is particularly sensitive to yellow/green, and hence picks out the center of the image circle rather distinctly compared to the red and blue surrounding it. To combat the remaining chromatic aberration, 17th century astronomers began building ever longer telescopes, using thinner objective lenses with longer focal lengths. This strategy worked because the color-blurred image remains the same size, while as noted in Fig. 5 the size of the image of an object grows in proportion to the focal length. Images that overlapped with a short focus could be separated enough to be distinct with a long focus.

Fig. 7. An engraving of the Paris Observatory during Cassini's time. The tower on the right is the "Marly Tower", moved there by Cassini for mounting long focus and aerial telescopes.

In the heyday of long telescopes, many were constructed as *aerial* telescopes with no tube, and a few reached preposterous lengths of 500 feet or longer, with goals such as seeing animals on the Moon. Cassini deployed 110-foot and 160-foot telescopes from Campani, and discovered differential rotation of Jupiter's atmosphere with the 160-foot, but the 34-foot model turned out to be the best compromise and by far the most successful. Figure 7 shows an engraving of the Paris Observatory with a number of long telescopes in use.[5]

---

[5] One hundred years later, lenses were fabricated from materials with different degrees of chromatic aberration that compensated each other and eliminated the overall aberration, allowing much shorter telescopes. It took so long because scientists (including Newton) mistakenly thought dispersion is proportional to the index and hence could not be canceled by using two materials. Also it is worth mentioning that telescopes with a reflecting mirror for the primary lens avoid the problem of refractive chromatic aberration entirely. Newton had built such a telescope but it had not reached the stage of maturity needed for such precise observations in the 17th century, though by the 20th century reflectors would surpass refractors for astronomy.

Operating such a long telescope as even the 34-foot must have been difficult, requiring a team of assistants. To measure the time of an emersion of Io once it had been established roughly when emersions took place, the telescope would most likely be pointed at Jupiter ahead of this time, and then continually adjusted to follow Jupiter as it moved across the sky with the Earth's rotation. The observer, let us say it was Picard that night, would then be watching the correct spot near Jupiter, while his team kept the telescope rotating with Jupiter, and at some point Io would begin to appear. Emersions of Io took about three minutes, and Picard would have decided from experience watching them what emersion stage he would use to fix the time; perhaps it was when he judged half of Io had emerged. Usually another Jovian moon would be nearby and could be used as a reference to judge the state of brightness of Io during an emersion or immersion. Variations in opacity of the sky or of background skylight would cancel in comparing the moons. (This method at least was used in the early 20th century to time eclipses, but we simply don't know what method was used at Paris in the 1670s; we do know now that it was very accurate — to 30 seconds or better.) Fortunately, the constant alignment of the telescope only had to be good enough to keep Jupiter and the moon somewhere in the field of view, so there was some slack in the alignment. Similar methods were used for immersions.

Now let us return to the night of November 9, 1676, where we left Picard and Rømer getting ready for the crucial attempt to verify the very public prediction of a delay in eclipse of Io. The Earth locations on November 9 and on the reference date of August 12 are sketched in Fig. 8. Apparently the date advertised was originally November 16. As it turned out this special observation was made on the evening of November 9, most likely as a hedge against its being cloudy on November 16 (as it probably was since there is no recorded observation for November 16). There must have been a serious concern though; the measurement would surely be more difficult on November 9 as the emersion would occur about 5 PM — though not the middle of the day, still two hours earlier in the evening than on November 16 and therefore viewed against a much brighter skylight background.

When the observations begin, Io is hiding in Jupiter's shadow and not yet visible. Someone must be keeping his eye on Huygens's pendulum

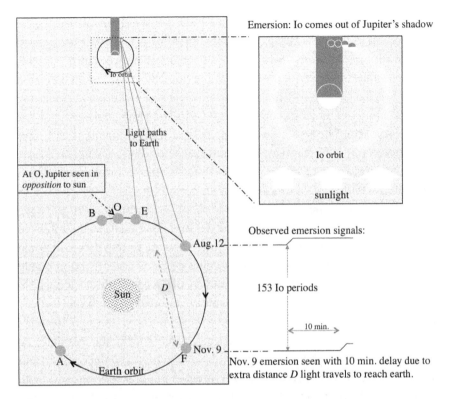

Fig. 8. Viewing Io immersions and emersions from the Earth in various positions in its orbit. At *E* the Earth has just passed *O*, its position of closest approach to Jupiter, called *opposition* since Jupiter and the Sun are in opposite directions from Earth. At such locations as *A* and *B*, when the Earth is approaching Jupiter, it is on the side of its orbit where immersions of Io (i.e. entering Jupiter's shadow) will be seen, while at *E* and *F* the Earth is receding from Jupiter and is on the side where emersions will be seen. Figure 2 is included here for clarity in picturing the emersion. The August 12 and November 9, 1676, locations are also shown with the predicted 10-minute delay on November 9 in emersion sketched on the right.

clock while Picard or Rømer looks intently at the spot in the telescope image near the brilliantly shining Jupiter where Io is expected to appear; it will start as a faint speck as it emerges from Jupiter's shadow and then grow to full brightness in about three minutes. During all this time the unwieldy telescope is continually adjusted to keep pace with Jupiter as the Earth rotates and maintain the scene in the field of view, while the telescope observations and clock times are constantly recorded, to be matched up at the end. Huygens's clock, good as it is, is still not a good enough time piece on its own; it serves to interpolate over short intervals between times of observed positions of the Sun and stars. For an emersion it only needed to keep accurate time from some well understood event, say the moment a reference star crosses the meridian, which depends upon the Earth's rotation. The real clock for the eclipse measurements is therefore the rotation of the Earth, which is far steadier for days, weeks, and years than man-made clocks of the period. Basically, the Io orbital period was compared with the regular Earth rotation period (the day).

As the astronomers wait, some of them — certainly Rømer — are expecting Io will be delayed by about 10 minutes based on the effect of the speed of light depicted in Fig. 8. They know of course they still could simply be wrong. The first crucial moment comes when emersion would begin if there were no delay compared to the August observations. No emersion is seen. As the minutes tick away, and they can tell that a delay is indeed taking place, thoughts could well be going through Rømer's head even as he concentrates on making observations; for example, he would be aware that the emersion light by now has passed the orbital point where the Earth was in August, and is streaking at this very moment with unimaginable speed across the extra distance to where the Earth has moved since then, but taking extra minutes to arrive at Earth because of the great size of the Earth's orbit. Finally, a faint speck near Jupiter does appear, marking the start of emersion (so they <u>can</u> see it in this twilight) and grows into the full brightness of Io. The time is noted, a time that was recorded and hence we know now showed a delay of several minutes as predicted. Two weeks later, with this public confirmation in hand plus several years of similar measurements, Rømer will be addressing the Academy and presenting his paper.

This is indeed a climactic point, though it marks just the start of a 50-year controversy. For such an important question the Academy held a special meeting a week later, and then scheduled another special meeting the following week at which Cassini would present his objections and Rømer his answers. Cassini, who early on may have supported Rømer's idea, now argued that if the idea were true, the other Jovian moons should show a similar effect which he claimed they did not. Rømer's formal response is lost, but in a note to Christiaan Huygens at the same time he explained his calculations in great detail, said that the data on the other Jovian moons were simply not yet accurate enough to see the delay, and emphasized that the effect depended upon the location of the *Earth* which should scarcely have a significant effect upon the motion of Io so far away but of course would be crucial in determining the distance light would travel to be seen on Earth. Huygens was totally convinced, and became a champion of *mora lumis*. Rømer then wrote the paper and undertook to spread the idea beyond Paris, especially to England where an excellent translation of his paper appeared. The eminent astronomer Edmund Halley took up the challenge to verify the idea further, with great success, and Newton soon became a convert, while in Paris there was staunch opposition, and little further work on it for decades.

By the end of the 1671–1678 Paris observations it was realized that there were extra variations of the Io eclipse period at the level of 20–30% of the speed of light delay; these variations could not be explained. These variations were the reason that Rømer's value for c turns out to be mistaken by about 25%. Eventually, as Rømer had argued, the effect of these on the observed delay should average away over time because they depend upon Jupiter and Io and should not be correlated with the Earth's movement about the Sun. Rømer returned to Denmark in 1678 and could not follow up on this argument. However, the English astronomer Edmund Halley, as noted already, took a keen interest in this question and indeed averaged Io eclipse observations in Paris and England over a 20-year period. In 1694, he published his results. He adopted as "most ingenious" Rømer's hypothesis of the finite velocity of light and presented an updated value for the speed of light based on the 20 years of observations, giving the time to travel the diameter of the Earth's orbit as about 17 minutes, yielding a value of c very close to the value we accept today.

It is interesting to contrast the views of Cassini and Halley. By the 1690s, Cassini had published tables for Jupiter's moons to be used for establishing the Longitude from observations of the moons anywhere on Earth. To account for the inequality in Io's eclipses, *which he does not wish to attribute to the finite speed of light*, he puts in by hand a 14 minutes 10 seconds earlier eclipse time when the Earth is nearest to Jupiter than when the Earth is furthest from it, to be interpolated for intermediate locations. In 1694, Halley published an adaptation for London of Cassini's new tables, praising their accuracy except for the part put in by hand.

Also, analyzing various observations of his own, at Paris he showed that the inequalities for the third and the fourth moons were not so large as claimed by Cassini, and were compatible with Rømer's hypothesis. In spite of his admiration for Cassini, Halley did not hesitate to strongly criticize his stubbornness in rejecting the idea of the finite velocity of light. In fact, Halley said: "there is good cause to believe that his motive thereto, is what he has thought not proper to discover." This indirect statement certainly has been interpreted as questioning Cassini's motives!

At the beginning of the 1700s, in England and elsewhere outside of France, we encounter more and more often the belief that light has a finite velocity, and that it was Rømer who proved it. Very influential were the favorable declarations made by Newton in his *Principia* and in his *Opticks*. Halley's thorough analysis, indicating a consistent value of the speed of light could be extracted after long periods of averaging, was quite convincing. Only in Paris, where the Cassini family remained in control of the Observatory, was this view not accepted, until 1729 when confirmation by a completely different method came from England as we will tell in Chapter 5.

However! No one at that time understood all the variations in the eclipse data; Halley and others just knew they averaged away over many years to give consistent light delays. And no one could know that there was a remarkable story hidden in the Paris data of the 1670s. We will take some time out to tell that story now. Then we will return to the concluding sections of this chapter.

Special Section: The Story Hidden in the 1670s Paris Eclipse Data

If the 25% size variations in the eclipse data are not due to large errors by Picard, Rømer *et al.*, then what is the cause? The fascinating explanation involves the linked motions of three of Jupiter's moons: Io, Europa, and Ganymede, which have orbital periods that are locked together close to the ratio of 1 : 2 : 4. Such orbital *resonances* are now known to be fairly common among moons in the Solar System. The three-moon resonance involving Io was the first one recognized, about 100 years after Rømer when it was

(Continued)

studied closely by the great French mathematician Laplace and by many others since then. According to these studies, the orbits of Io and Europa gradually changed over billions of years until the orbital period of Europa reached very close to twice the period of Io at which point the effect of their mutual gravitational attraction was synchronously enhanced and pulled them into slightly stretched (elliptical) orbits where they have since remained locked in resonance, and the orbits have evolved together. (A similar thing happened between Europa and Ganymede.) The upshot is that Io has an oblong orbit that makes a complete rotation in 400 days. This rotation will have a small effect on when Io enters and leaves Jupiter's shadow and thus will change the observed eclipse times. Halley obtained an average light delay which we now know is quite accurate. He just didn't know the reason for the variations about this average.

Although neither Laplace nor anyone else knew it, his prediction had already been verified a century earlier in the Paris observations. *The 400-day rotation of Io's elliptical orbit causes a change in the Io eclipse times (including the time to catch up with Jupiter's rotating shadow) of every 440 days.* This is the effect that sat quietly in the Paris data for centuries. The residual scatter of the Paris eclipse times (after the light propagation delay, Jupiter shadow rotation, and all other known effects have been removed) is not merely a jumble, but instead forms a pattern that is just what would be expected due to the rotation of Io's orbit, in complete accord with Laplace's 18th century analysis, and with the most recent telescope and spacecraft measurements.

This oblong orbit manifested in the Io eclipse data also profoundly affects the surface of Io. We can see this by first looking at our own Moon, which spins about its axis with the same period as it revolves in its orbit about the Earth, thereby always keeping the same face of the "man in the Moon" pointing toward us on the Earth. This came to be because, just as the Moon raises tides on the Earth, the Earth has raised tides on the Moon (even when the Moon became solid there would be tidal deformations) which over billions of years gradually slowed down the Moon's spin by friction until at last the spin matched the orbit and the tidal deformation did not move and dissipate energy. A similar effect took place with Io except the oblong

(Continued)

(Continued)

elliptical orbit with its changing angular speed (which is the cause of the small eclipse variations!) prevents a perfect match between spin and orbit, and keeps the tidal distortion due to Jupiter from being perfectly stationary on Io. Likewise for Europa. Jupiter is so massive that its tidal dissipation on these moons would have circularized the orbits, permitting a perfect spin–orbit match as with our Moon, *except* the oblong elliptical shape is locked in by the three-moon resonance. The effect on the surfaces of these moons is profound, as revealed by NASA missions in the past few decades. The tidal forces are strongest on Io, the closest to Jupiter, heating up the moon and evaporating all water, while generating internal heat to constantly power volcanoes that dot the surface. On Europa the tidal forces are weaker but still generate sufficient heat to maintain melted water beneath the ice cover, in fact an ocean 60 miles deep, giving rise to much recent speculation about the possibility of life there.

After all the corrections the magnitude of residual scatter in the Paris data is only about 30 seconds, which provides a measure of the errors in the Paris measurements of the eclipse times. This error is impressive; the combination of observational acuity with this long telescope, coordinating with Huygens's pendulum clock, and interpolating between timed positions of the stars as the Earth rotates, is remarkable. *These were great astronomers*! Let us now go back and follow the highlights of their remaining lives to conclude our story.

We start with Cassini. We can agree with Halley's criticisms of Cassini in the one instance of the speed of light and still acknowledge Cassini's great accomplishments. Longitude is a good example. Having mapped France earlier, Cassini moved on to map the world. He trained clever young Jesuit priests assigned to follow French explorers, who then sent Latitude and Longitude measurements back to Paris. Cassini laid out a 24-foot circular map on the floor of the Paris Observatory tower, with the north pole at the center. As Longitude values came in from around the world, he added such cities and farflung locales as Quebec, Santiago, Lisbon, Venice, Cairo, Siam, India, Canton, and Peking. The first accurate map of the world slowly took shape. In 1696, Cassini published his new map, the most accurate of the time. Cassini took this method of Longitude about as far as it could go, given its practical problems; for example, it was not useful for ships at sea because precise telescopic readings were too difficult from a heaving ship platform. A practical way to determine Longitude had to wait for the development of exquisite mechanical clocks in the 18th century and at about the same time a lunar motion clock. As an indicator of how seriously the Longitude problem was viewed, King Charles of England created the Royal Observatory at Greenwich and the position of Astronomer Royal to develop the lunar motion clock. (The entire fascinating story is told in the book *Longitude* by Dava Sobel.) Cassini was a gifted observer, and was most famous for his work on the planet Saturn. Today, the remarkably successful Saturn space mission is called Cassini–Huygens; the spacecraft is named after Cassini, and its probe that explored Titan honors Huygens.

Picard died in 1682 with a secure reputation as one of the great observational astronomers — perhaps the greatest of his day. The beauty of the 1671–1678 eclipse measurements, for which he should be given ultimate credit, was perhaps not appreciated until recently, but his other work, only touched upon in these pages, insured his lasting reputation. It has never been clear what stand Picard took in the debate between Rømer and Cassini, although there is some implication in Rømer's letters to Huygens that Picard was at least somewhat skeptical of Rømer's conclusions. Picard also may have simply shunned controversy.

Among Picard's noteworthy achievements was the first accurate measurement of the Sun's diameter, especially important as they were made during a period when the solar activity was minimum and the Sun nearly without sunspots. It is fitting that a new satellite mission to study the Sun and its effect on the climate of the Earth is named PICARD and was launched on June 15, 2010.

Now, what about Rømer? He returned to Denmark in 1678 where he was appointed professor of astronomy at the University of Copenhagen and the same year married Anne Marie Bartholin whom he had known, as mentioned already, when living in Erasmus Bartholin's house.

In Paris Rømer had shown his remarkable aptitude for large ideas and at the same time his inventive genius with machines and gears. In Denmark he performed at both levels for the rest of his life. He remained active in astronomy, using improved instruments of his own construction such as the world's first meridian circle for locating and timing the positions of stars more precisely than ever before. His circle was eventually adopted throughout Europe with major impact on observational astronomy.

In 1683, Rømer was appointed Royal Mathematician by the king; in this position he started the first national system for weights and measures in Denmark. He introduced the first modern thermometer. His innovative temperature scale was based on two fixed points, boiling and freezing, and is still known as the Rømer scale (not to be confused with the later Reaumur scale). The idea, so familiar to us now, of using two fiduciary points with equally spaced calibration marks between them was completely new. Rømer enclosed a liquid in a sealed glass tube which made it immune to pressure changes. The liquid was an improvement, too, a

Fig. 9.   Ole Christensen Rømer dressed in full regalia, befitting his position as Mayor of Copenhagen.

mixture of alcohol and water familiar in the form of wine — but doubtless not used in this form! This mixture avoided the drawbacks of both the low boiling point of pure alcohol and the odd behavior of pure water near freezing. Daniel Gabriel Fahrenheit learned of Rømer's work and visited him in 1708; in one of his letters Fahrenheit narrates how he borrowed the idea for the scale from this visit, eventually establishing what is now known as the Fahrenheit scale, in 1724. The Celsius scale came shortly thereafter.

In addition to his remarkable scientific career, Rømer also served in important public positions. He was appointed Chief of the Copenhagen Police and then became Mayor of Copenhagen. His portrait in this position is shown in Fig. 9. His accomplishments in these offices are well known in his native land. With such a varied, productive life, one can only wonder whether Rømer felt that the discovery of the velocity of light was still his greatest accomplishment. Perhaps he felt some sadness that it was not accepted during his lifetime in its birthplace — Paris — and yet the awareness that it had happened when he had the opportunity to learn in and be a key part of a great international team of astronomers assembled there, being the greatest such gathering in one place for a very long time to come.

Chapter 5

Speed of Light 2: The Advent of Precision Astronomy

"It was chiefly therefore curiosity that tempted me to prepare for observing the star on December 17th, when having adjusted the instrument as usual, I perceived that it passed a little more southerly this day than when it was observed before."

James Bradley, in a 1725 letter to Edmond Halley

In December 1725, James Bradley, a church minister turned Oxford Astronomer, and gentleman scientist Samuel Molyneux together began an attempt to make the first detection of *parallax*, the shift in direction of a distant star seen from Earth as the Earth moves in its orbit about the Sun. Seeing parallax would be observational proof of the Earth's orbital motion, and the size of parallax would reveal the distance to the stars, then completely unknown. Molyneux's telescope, exquisitely designed for just this purpose, was located on his property at Kew in London. It was to mark the beginning of a new level of precision in astronomy, a revolution in the accuracy of locating the relative positions of stars (and planets) in the heavens. With Bradley away in Oxford, Molyneux began pinpointing the exact location of their selected star Gamma Draconis on the nights of December 3, 5, 11, and 12. Their telescope could detect minute northerly or southerly changes in the location of this star. It was most sensitive to daily changes in parallax during March and September and quite insensitive during December and June. Bradley arrived at Kew on December 17, and Molyneux's observations having shown no evidence of a shift in position, just as anticipated, Bradley then states: "… farther repetition of them at this season seemed needless, it being a part of the year wherein no sensible alteration of parallax in this star could soon be expected. It was chiefly therefore curiosity that tempted me to prepare for observing the star on December 17th, when having adjusted the instrument as usual, I perceived that it passed a little more southerly this day than when it was observed before."

A surprise. And on subsequent days, painstaking measurements showed that the star continued to move southerly, all the more surprising since, as we will see later, any small change in position due to parallax should have been in the *northerly* direction. Bradley and Molyneux therefore well understood this could not be parallax, but then what was it? At that point, they just didn't know. Little did they suspect that their observations were to provide a new method of measuring the speed of light that would confirm the speed obtained in Chapter 4 from the eclipses of Jupiter's moon Io, and put completely to rest any remaining doubts, in Paris or elsewhere. But none of this happened right away. It was to be 3 years — and after the sudden death of Molyneux — before Bradley came up with the explanation of their observed motion. *Bradley said that if light has a finite speed then light from a star would appear to come from different directions as the earth changed its direction of motion around the sun.* The change with motion was given the fancy name of *stellar aberration* (as distinct from the *chromatic* aberration of lenses discussed in the last chapter), and Bradley's explanation was immediately accepted by scientists of the day. We will discover Bradley's explanation for ourselves in due course by the same reasoning he used. It was the first direct demonstration that the Earth indeed moves around the Sun, as definitive as detection of parallax would have been, though knowing the distance to the stars would still have to wait 100 years until the detection of parallax itself. And then knowing how truly large the universe is would have to await the discovery of stars in the early 20th century by Henrietta Leavitt that were called *standard candles*, as explained at the end of this chapter.

To begin this story, we must go back to well before Bradley and Molyneux to the same time-frame as the Paris observations in the last chapter, the 1660s and 1670s. At that time, not only was the practical topic of Longitude important to astronomers, but the purely scientific topic of parallax was also very much under discussion. The Copernican view of the Earth orbiting the Sun implied that the "fixed" stars outside the solar system should not appear fixed but instead should exhibit parallax. As illustrated in Fig. 1, the closer the star, the larger the parallax, much as nearby trees seen from a speeding 1660s coach would rush by faster than distant ones. The failure to detect parallax was construed as evidence against the Copernican system by many, since the stars would have to be so far away as to be inconceivable to them. For the majority of the experts, however, the success of Kepler and Newton in explaining the orbits of the planets about the Sun overwhelmingly settled the question in favor of the Copernican view. Still, they thought it of transcendent importance to

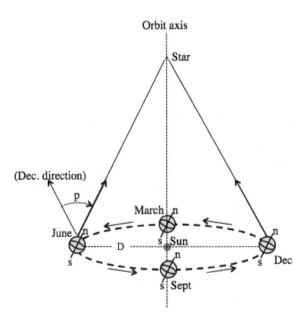

Fig. 1.   Parallax, the change in angle of a distant object when viewed from two ends of a baseline, is shown here for a distant star using the Earth orbital diameter D as the baseline. For simplicity the star is taken to lie along the center axis of the orbit. The parallax angle p indicated here is the change in direction of the star in June compared to December, when viewed from a latitude near London. Knowing p and the length of the baseline then yields the distance to the object by trigonometry.

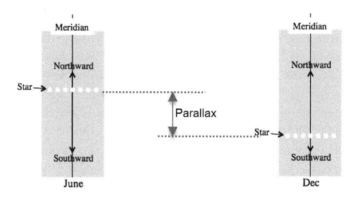

Fig. 2. How a single star viewed from the ground can reveal its parallax by its location when it crosses the Meridian, the North-South line in the sky that passes through the zenith. Inspection of Fig. 1 shows that the star will make its daily overhead passage more to the north in June than in December.

detect stellar parallax and with it to find out, by triangulation, the distance to the stars.

One way parallax might show up in actual observation of the stars is illustrated in Fig. 2, and is the starting point for all experiments discussed in this chapter. Robert Hooke took up the challenge of stellar parallax. Hooke was widely recognized by contemporaries as one of the great scientists of his era, although also well known for his continual public disputes. He was born in 1635 on the Isle of Wight into a family of ministers and confirmed royalists, a political persuasion that would prove beneficial to him.

Hooke, a frail child with physical deformities that plagued him throughout his life, entered Oxford in 1653, where he maintained the family tradition as a staunch monarchist. The versatile Hooke supported himself initially by securing a chorister's place at Christ Church after twenty lessons on the organ. At this time, added to his childhood ailments (headaches, stomach upsets, sleeplessness), Hooke's posture had begun to deteriorate to the point that his body was bent over from curvature of the spine. Still he gained employment as an assistant to Robert Boyle and constructed, operated, and demonstrated Boyle's air pump, the *machina Boyleana*. He played a key role in obtaining *Boyle's Law*, which stated that the volume of a gas is inversely proportional to the pressure confining it.

Hooke was greatly admired for his ability and public spirit in the recovery from the two terrible calamities of the 1660s, the Great Plague of London in 1665, and a year later the Great Fire of London which consumed over 13,000 houses and 80 churches, including St. Paul's Cathedral. Among his civic contributions, Hooke became a Surveyor to the City of London, and appears to have performed more than half of all the surveys after the fire.

It had proved pivotal in Hooke's scientific career that upon the Restoration, the Royal Society of England was founded at Gresham College in London with a royal charter from Charles II. The founders, royalists with a close connection to Charles, included Boyle and another of Hooke's close acquaintances, Christopher Wren, later to become England's most famous architect and designer of the new St. Paul's Cathedral. The youthful Hooke had so distinguished himself that he was appointed the first Curator of Experiments, and later elected to Society membership. Hooke was granted an annuity, which gave him a reasonably comfortable livelihood and enabled him to pursue scientific interests the rest of his life and to acquire permanent lodgings in the Royal Society building at Gresham College, where he was a lecturer in Astronomy.

Thus, in 1668 Hooke could undertake his experiment to detect Parallax. His concept was remarkably well thought out and its implementation was clever, but the mounting of his telescope was not at all stable enough for the precision he would need to see something significant. However, he had the great insight to select a reasonably bright star, Gamma Draconis in the constellation Drago, the dragon, which was very close to the zenith (i.e. located straight up) at London when it crossed the meridian. This solved a major potential problem, the bending of starlight by refraction of the atmosphere, which could be mistaken for the effect of parallax. Just as a vertical beam of light remains unbent and vertical by symmetry after it enters the horizontal surface of a pond, starlight entering the atmosphere from straight above will be bent very little if at all. An additional advantage of the zenith direction is that the starlight will be closely aligned with a vertical plumb line, providing a convenient reference to see small deviations in starlight direction due to parallax.

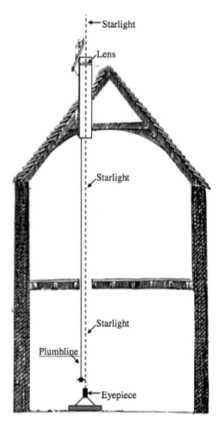

Fig. 3.   *The Archimedean Engine* is an allusion to Archimedes moving the Earth with his lever; it is the name Robert Hooke gave the experiment mounted in his apartment to see the movement of the Earth in its orbit by detecting the resulting parallax of the near-zenith star Gamma Draconis. The vertical light path from lens to eyepiece was 36-foot. Two plumblines (one is shown) defined the zenith direction. The same basic layout was used 50 years later by Bradley and Molyneux in their zenith telescope, but mounted with much greater stability and with a vertical tube surrounding the optical path from lens to eyepiece.

Hooke installed a 36-foot telescope (sufficiently long to overcome the chromatic aberration discussed in Chapter 4) in his apartment at Gresham, with holes in the ceilings and roof to allow starlight to be observed from the ground floor (see Fig. 3).

He suspended plumb lines next to the line of sight to locate stars relative to the zenith. The telescope consisted only of the objective lens at the top and the eyepiece at the bottom with no connecting tube to guard against perturbing air currents in the optical path. Hooke could look at the stars through the eyepiece while lying on a couch. He could see Gamma Draconis well enough day or night, and when it came into view, Hooke would measure the angle north or south of the zenith where the star crossed the meridian as in Fig. 2. To detect parallax Hooke would look for a change in this angle as the Earth moved in its orbit about the Sun (Figs. 1 and 2).

From the beginning, Hooke encountered serious problems with vibrations and instabilities, which could not be overcome without much more careful construction. Still, he thought he had seen the effect of parallax in measurements he had taken six months apart, and elected to announce the discovery in 1674. The scientific community had grave doubts about the validity of Hooke's discovery, and in fact Hooke stopped making his claim (though not renouncing it either). He never undertook the measurement again. Thus, in one sense, Hooke's experiment was a failure; in another, however, it was brilliantly conceived, and the ideas behind it impressed and inspired Bradley and Molyneux 50 years later. They modeled their own experiment after Hooke's, including his choice of star, Gamma Draconis, but used far more care than the temperamental and impatient Hooke.

After Hooke, the question of parallax lay dormant for over 50 years until Samuel Molyneux launched a new attempt at it in 1725. Samuel Molyneux came by his wealth and interest in science naturally. His father, William Molyneux, was an Irish landowner with a keen interest in optics and optical instruments whose major contribution was his 300-page book, *Dioptrica Nova*, the first optics book in English. When William died in 1698, his son Samuel was nine years old, and was then brought up by his uncle Thomas. At 16, Samuel entered Trinity College, Dublin, and after many adventures was elected to the London Parliament, and in 1717 married Lady Elizabeth Diana Capel, the eldest daughter of the Earl of Essex. Before this union, Molyneux had a comfortable financial position, but now he was wealthy. On top of his other activities, he renewed his interest in science and became active in astronomy and optics. He met James Bradley, like Molyneux, a fellow of the Royal Society and they shared an active interest in building affordable telescopes so a far wider range of

Fig. 4. Kew Palace about the time Molyneux had the zenith telescope installed there. The three stories could readily accommodate the 24-foot vertical tube though it might have seemed incongruous in this regal estate to have holes in floors and roof with a telescope poking through. The installation might well have been an object of curiosity, perhaps even fashionable to visit, but of course closed to the public whenever Gamma Draconis was overhead.

people could become astronomical observers. Also in 1721, on the death of Lady Capel of Tewkesbury, a relation of Molyneux's wife, the Molyneuxs inherited Kew Palace shown in Fig. 4, and Molyneux set up an observatory in the palace house there.

All along Molyneux had felt an abiding interest in resolving the still unsettled Parallax question, and now he could do something about it. He commissioned George Graham, a celebrated Fleet Street craftsman, to build a 24-foot. telescope, which was mounted in Molyneux's palace at Kew in November 1725. Molyneux had already approached his friend James Bradley to collaborate on the observations. He could not have made a better choice.

James Bradley was born in 1693 at Sherborne, Gloucestershire, to William Bradley and Jane Pound, aristocrats of decent means but not wealthy. He often stayed with his uncle the Rev. James Pound, a vicar at Wanstead in outer London, whom he idolized. Rev. Pound was an expert amateur astronomer and friend of Newton and Halley. In 1717, the Royal Society lent Huygens's 123-foot focal length glass objective (made as described in Chapter 1) to Rev. Pound for a long aerial telescope to be set up in Wanstead Park near the rectory. Newton and Halley utilized Pound's measurements, in which Bradley assisted and built a reputation of his own. Bradley had attended Balliol College, Oxford, and taken his MA in 1717, the year he contracted smallpox. He recovered under his aunt and uncle's care at Wanstead while also making observations with his uncle. He then followed his uncle into the clergy, becoming vicar of Bridstow near Hereford in 1718. At the same time, he was elected into the Royal Society. In addition, his new acquaintance, Samuel Molyneux, procured for Bradley a small sinecure in Wales. In 1721 the Savilian Chair of Astronomy at Oxford came vacant and, with Newton's recommendation, Bradley at the early age of 28 received this prestigious appointment and left Bridstow to take up astronomy full-time.

From this point on, Bradley devoted his energies almost exclusively to astronomy in contrast to both Hooke and Molyneux, who spread their energies broadly over diverse scientific interests. Bradley had immediately begun to fulfill the high promise expected of him, and in 1722 measured the diameter of Venus with another large aerial telescope. In 1725, at age 32, he joined Molyneux to use the new telescope Molyneux was having installed at Kew.

Construction began by boring holes through the ceiling and roof of Molyneux's house as was done in Hooke's house 50 years earlier. But there the similarity ended. Although the basic scheme was similar to Hooke's, the implementation was much sturdier. The 4-inch wide objective lens and the eyepiece were fixed at either end of a 24-foot vertical tube formed from strong, firm tin plate. As shown in Fig. 5, the telescope could be rotated about an axis near its top to point north or south of the zenith. This assembly was supported by iron-work fixed strongly, in Molyneux's words, to a "large stack of brick chimneys, which were quite within the house, and scarce at all exposed to the weather, and were very strong old built chimneys, some part of the house being near three hundred years old." *Thus a house with sturdy chimneys provided the stability for the new precision needed for observations, and became the forerunner of equally sturdy observatories built exclusively for telescopes.*

To Star

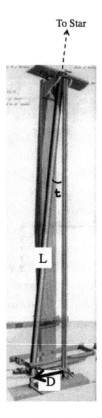

Fig. 5. Sketch of the zenith telescope at Kew, length $L = 24$ ft. To center the star in the eyepiece when it crossed the meridian, the tilt angle t was adjusted relative to the plumbline by fine-tuning the displacement D with a micrometer.

With the upper end of the tube thus secured to the iron pivot, the location of Gamma Draconis as it crossed the meridian was determined by rotating the lower end (containing the eyepiece) along a north–south arc until the tube pointed to the star, and then fine-tuning the direction with a micrometer screw. When Gamma Draconis was centered relative to fine wire cross hairs in the eyepiece, the exact distance D, and hence the tilt of the telescope, was read off a micrometer scale whose zero-point was indicated by a plumb line hanging from the top of the telescope. Starting December 3, 1724, Molyneux recorded the first "official" tilt reading of Gamma Draconis when it crossed the meridian line as in Fig. 6(a), and the

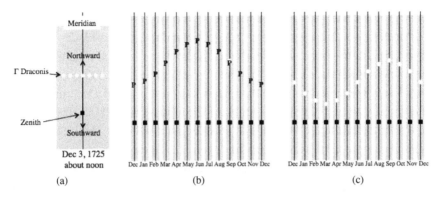

Fig. 6. Gamma Draconis crossing the meridian. (a) Sequence of "snapshots" about every minute on December 3, 1725, pinpointing the crossing point. (b) *Expected* crossing points each month if north/south displacements are due to parallax as in Figs. 1 and 2. Slowest change per month is in December and June; fastest in March and September. Vertical scale arbitrary. (c) *Actual* displacements observed by Bradley and Molyneux December 1725–1726. Slowest change per month is in March and September; fastest in December and June. Vertical scale is about 40 arcsec between displacement maximum and minimum.

north/south deviations on subsequent days were then easily measured by the micrometer screw.

Bradley and Molyneux recorded their tilt-angle readings in seconds of arc (sometimes called arcsec which is truly small). To change the tilt of the 24-foot vertical telescope *t* by one arcsec, the bottom was moved horizontally (changing *D* in Fig. 5) by 1.4 thousandth of an inch, tiny but big enough to read with a micrometer screw. So how large was the image spot of Gamma Draconis in this telescope? The 4-inch objective lens in this scope would have a spot size due to chromatic aberration of 16 thousandths of an inch, corresponding to a change in tilt angle of about 11 arcsec. In their stable system, the location of the spot center could be determined to better than 10% of its diameter, i.e. one arcsec or less.

By contrast with Robert Hooke's telescope, George Graham's was remarkably stable, taking every advantage of the solid citadel to which it was attached; it scarcely moved one or two seconds of arc in the strongest winds. Molyneux's notebook contains many colorful commentaries on extreme weather that hardly budged the scope's orientation. For example: "On Tuesday December 7 there was in the morning a very violent and unusual hurricane, such as hath not been known in many years, and from that time to this day a very great change in weather, from cold, dry frost, to mild, open, rainy, warm weather. Notwithstanding all which changes, when the mark was adjusted this day again to the plumbline, the index which was found at 9.8 [arcseconds] as it had been left on the 5th of December, was altered only to 11.5 so that the stack of the chimneys, with all this change of weather, had altered but about a second and a half, by which the top of them was come northward." And then again: "… after a very violent stormy rainy night, Mr. Graham examined the instrument very carefully …it had altered since the day before but about half a second, by which the tip of the chimneys had yielded southward." And again, on December 12: "After an exceeding tempestuous rainy night, Mr. Graham again examined the instrument, and adjusting again the mark to the plumbline … he found the index stand at 10.2 whereas the day before it had stood at 11, so that it had not altered one whole second in that time, by which the top of the chimneys were come southward." (One can only imagine how Hooke's telescope would have responded on such occasions!) Note that only small adjustments were needed to realign with the ultimate reference, the plumb line, which likewise must have held quite steady. A revolution in precision had begun, though not complete and not widely known as yet.

Bradley and Molyneux expected any motion of Gamma Draconis due to parallax to be similar to that shown in Fig. 6(b). However, when we left them just after the early paragraphs of this chapter, they were seeing quite a different motion. Let us now pick up where we left off. While the two astronomers made exhaustive checks for possible instrument defects, and finding none as an explanation, the star was seen to reach its maximum southward excursion in March (rather than December), then start to head northerly, returning to the first position in June and reaching its maximum northward extent in September, finally returning to the original place and heading south again in December. This actual motion during the first year is shown in Fig. 6(c). What was happening? Here were annual changes in position as expected with parallax, but clearly not due to parallax because the maximum displacement (Fig. 6(c)) occurred when parallax should have yielded a minimum (Fig. 6(b)), and vice versa.

Both Bradley's and Molyneux's notebooks are filled with hypotheses as to the cause of the unexpected observed motion, plus discussions and rejections of most ideas. They had, however, obtained a lead by observing the few stars besides Gamma Draconis that their telescope, with just a very narrow field of view (FOV) around the zenith, allowed them to see. These few stars moved in step with Gamma Draconis. To follow up this lead adequately, Bradley and Molyneux eventually concluded they should keep tabs on many more stars and hence would need a much larger FOV than afforded by the telescope at Kew. Bradley decided to build a second telescope rather than remodel the one at Kew.

Bradley located the new telescope at his familiar observing station at Wanstead in the house of his late uncle James Pound, another stable old dwelling. This location had the advantage that Bradley could lodge on site, rather than travel to Kew to make observations and return to sleep, as he had been doing. George Graham was again commissioned to build this version, using very much the same design. However, to allow them to see more stars north and south of zenith, Graham constructed a longer arc sector about 3 feet in extent, along which the bottom of the telescope tube with the eyepiece could be swung. The tube itself had half the length of that at Kew, 12 feet, constrained by the dimensions of the smaller house. Although the resulting angular resolution would be about a factor of two coarser than at Kew, it would still be quite adequate. Moreover, this telescope could swing through a 100 times wider arc giving it the desired greater FOV with equally good stability and small enough atmospheric bending. The telescope has been preserved very well, and is on exhibit at the Greenwich observatory. Bradley's aunt gave him permission to cut holes in the roof and floor next to the chimney as had been done at Kew. However, this time the house had just one story, and even with the shorter tube the eyepiece had to be below the main floor, so that Bradley made his observations from the coal cellar.

The Wanstead telescope was erected in August 1727, with Molyneux assisting despite having been appointed one of the lords of the admiralty, and Bradley began thorough tests of the instrument, in part based on issues he had encountered at Kew. Among many concerns, Bradley worried about changes in the environmental conditions, such as temperature and barometric pressure. His concerns were well founded. The temperature, for example, could be different over the course of a year depending upon the season and the time of day or night the star appeared. Changes in temperature would expand or contract the length of the arc along which Bradley measured the position of the tube when sighting a star. The resulting change in the position of the eyepiece relative to the reference plumb line could introduce an error in the location of the star. We can make a simple estimate of the magnitude of the problem. A change in temperature of 1°C causes brass to change its length by 20 parts per million. Thus, there would be a change in tilt angle of 1 arcsec per °C at the endpoint of the 3-foot arc Bradley had installed at Wanstead. Clearly, a few degrees change in temperature could be a major source of error at the 1 arcsec level of accuracy Bradley was reaching. At Wanstead, Bradley instituted the procedure of recording the temperature and pressure for every observation, which would help set the limits of accuracy. Bradley was the first astronomer to take such precautions, and he set a standard for precision measurements that became a major influence on later experiments in astronomy and other physical sciences.

Using the new telescope Bradley could view about 200 relatively bright stars near Gamma Draconis, many more than at Kew. Bradley took painstaking observations of all these stars, and found that each moved with the seasons approximately in step with Gamma Draconis. Important information, but still offering no answer to the simple question from the earliest data at Kew: Why do any stars move this way? Indeed, since it's not parallax, why do they move at all?

In the midst of these developments, a great tragedy occurred. In 1728, Molyneux suffered a *fit* (a seizure or convulsion of some kind) while in the House of Commons. He was treated by court anatomist Nathaniel St. André, but after a few weeks Molyneux died. On the night of the death, St. André eloped with Molyneux's wife, Elizabeth, the two marrying in 1730. Samuel Madden, a relative of Molyneux's, claimed that St. André had poisoned the MP. Although St. André won an action for defamation, he was unable to secure regular work, and he and Elizabeth lived rather isolated from then on. Added to the sorrow of Molyneux's death is the sad irony that he never found out the answer to the great riddle of Gamma Draconis, but died just shortly before Bradley had a profound insight and figured it out.

Bradley's moment of enlightenment in the summer of 1728 has been recorded by Thomas Thomson in his 1812 *History of the Royal Society*:

"At last, when [Dr. Bradley] had despaired of being able to account for the phenomena which he had observed, a satisfactory explanation of it occurred to him all at once, when he was not in search of it. He accompanied a pleasure party in a sail upon the river Thames. The boat in which they were was provided with a mast, which had a vane at the top of it. It blew a moderate wind, and the party sailed up and down the river for a considerable time. Dr. Bradley remarked, that every time the boat put about, the vane at the top of the boat's mast shifted a little, as if there had been a slight change in the direction of the wind. He observed this three or four times without speaking; at last he mentioned it to the sailors, and expressed his surprise that the wind should shift so regularly every time they put about. The sailors told him that the wind had not shifted, but that the apparent change was owing to the change in the direction of the boat, and assured him that the same thing invariably happened in all cases. This accidental observation led him to conclude that the phenomenon which had puzzled him so much was owing to the combined motion of light and of the earth."

As shown in Fig. 7, the vane points more toward the rear of the boat when the boat is moving, indicating the wind appears to be coming more from the forward direction; if the boat turns around, then the wind will appear to shift and come more from the new forward direction. By analogy, if the speed of the wind is thought of as the speed of the starlight, and the speed of the boat as the speed of the Earth in its orbit, then the vane direction should be thought of as the direction of the starlight seen from the moving Earth, which is measured by the tilt of the telescope centered on the star, as in Fig. 5. Bradley reasoned that the starlight likewise would appear to come more from the forward direction of the Earth's movement about the Sun, and would shift a half year later when the Earth moves in the opposite direction. *To help him conceive of this idea, how fortunate that Bradley was well aware of Rømer's discovery about the speed of light, and in fact had investigated some of the issues with Jupiter's moons himself. Thus, Bradley was conceptually primed for his revelation.*

It's entirely possible that the sailboat account was thought up by someone as a nice story years later since there is no contemporary account in Bradley's journals or elsewhere. But the story does seem plausible, and

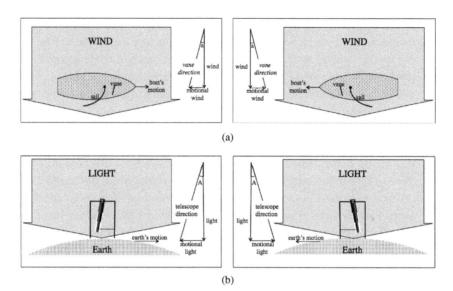

(a)

(b)

Fig. 7. Bradley's revelation. (a) When a sailboat moves in an external wind (assumed a side wind here for simplicity) the wind vane will point in the direction of the external wind *combined with* the backward streaming motional wind induced by the forward motion of the sailboat. The vane thus changes direction by twice the angle *a* shown when the sailboat reverses its heading. (b) By analogy, when the Earth moves in an external beam of light from a star, the light seen by a telescope on the Earth has the direction of the external light combined with the back streaming motional light induced by the motion of the Earth. When the Earth in orbit about the Sun reverses its direction of motion, the direction of light from a star will appear to change by the angle 2A, and a telescope must be realigned by the angle 2A to view the star, and is thus a "light vane".

has been accepted as true for two centuries. In any event, the sailboat provides a good analogy for the effect of stellar aberration.

Another good analogy for stellar aberration is tilting an umbrella in the forward direction to block the rain when running in a downpour as shown in Fig. 8(a). Figure 8(b) shows the change in umbrella tilt to point in the apparent direction of the rain during a dash around a circular track in a downpour. We can imagine a similar reversal of tilt of a telescope pointed in the apparent direction of the light from Gamma Draconis as the Earth makes its dash around the Sun.

Bradley published his result in the Proceedings of the Royal Society in January, 1729. He included only data taken at Wanstead in order, he

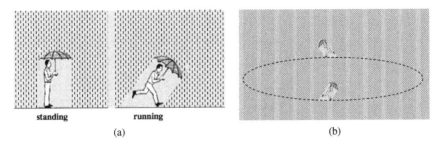

Fig. 8. Showing how the direction of falling rain or "falling" light would be altered by an observer's motion. (a) When running in a downpour the umbrella is tilted in the forward direction to block the rain. (b) Note how the tilt reverses during a dash around a circular track. Likewise, the tilt of a telescope must reverse to receive the light from a star as the Earth makes its dash around the Sun.

said, to provide internal consistency using one telescope and one observer. Nevertheless, the observations at Kew were, of course, absolutely crucial. Not only did they give results for Gamma Draconis that were very similar to the later observations at Wanstead, but they were part of a learning process that led to the good practices at Wanstead such as recording temperature and pressure. Moreover, had Molyneux and Bradley not undertaken the observations at Kew that had revealed the mysterious new phenomenon, the telescope at Wanstead likely would not even have been built.

Bradley emerges as probably the preeminent astronomer of his generation in all Europe. His published data from Wanstead are shown in Fig. 9. They are plotted together with the curve based on his great insight, the idea of stellar aberration, with the height of the curve adjusted to the best match with his data points. (The shape of the curve follows the component of the Earth's orbital velocity that lies along the north/south meridian at Wanstead, which changes with the months as seen in Fig. 9 due to the motion of the Earth in its orbit. For the readers familiar with trigonometry, this curve is close to a simple cosine function of time with a period of one year.) The curve based on Bradley's insight is an excellent fit. The 6 arcsec height of the error bars in the figure shows Bradley's estimated error of ±3 arcsec for each measurement, and is about 15% of the size of the 22 arcsec image spot from chromatic aberration for the Wanstead

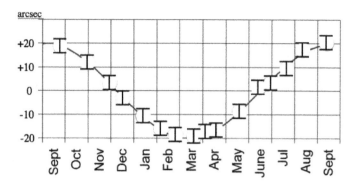

Fig. 9.   Bradley's 1727–1728 data on the north–south movement of Γ-Draconis establishing stellar aberration. Also plotted is a curve having the shape expected from aberration (the same as given by the sailboat or umbrella analogies), with the height adjusted to the best match with the displacements shown by the data points.

telescope. (We estimated 11 arcsec for this spot size in the Kew telescope, and the Wanstead telescope image spot should be twice as big since the scope is half as long.) Bradley specified aberration, the change in angle of the telescope when the Earth moves at right angles to the direction of the star, by the angle A shown in Fig. 7(b). To find A, Bradley needed only a small correction to the values shown in Fig. 9 for September and March. Bradley's calculated value A, 20.5 arcseconds, is quite remarkable since it is almost precisely the modern value of 20.47 when corrected back to the average speed of the Earth in Bradley's time. Clearly, Bradley was a gifted observer!

Bradley realized he could use his data to obtain the speed of light. We show in the numerical side-bar below that his measurements indicate that the speed of light must be about 185,000 mi/sec. (The modern value is 186,000 mi/sec.) Actually, Bradley compared his results directly with measurements of the eclipses of Jupiter's moons, avoiding the uncertainty that was then obtained in the size of the earth's orbit in terrestrial units, e.g. miles, and found the time for light to travel the diameter of the earth's orbit to be 16 min 40 sec. This value is very close to Halley's average value of 17 min from Io eclipses, cited in Chapter 1. It was very convincing that these two completely different methods gave values in such close agreement with each other.

Numerical sidebar: Bradley's aberration angle is A = 20.5 arcsec. We want to interpret this 20.5 arcsec tilt of the telescope in terms of the speed of the Earth v and the speed of light c. Since we found earlier that a tilt of 1 arcsec is a displacement (defined in Fig. 5) of D = 4.86 millionths the length L of the telescope, 20.5 arcsec means to good accuracy, D/L = 1/10,000. But from Fig. 7(b), $v/c = D/L$ so $v = c/10,000$. Today we know the speed of the Earth is about v = 18.5 mi/sec, so Bradley's aberration result implies c = 185,000 mi/sec. To compare instead with the eclipses of Jupiter's moons, as Bradley did, we use time = distance/speed: the time t for light to travel the diameter $2R$ of the earth's orbit is $t = 2R/c$, while the time for the Earth to travel the circumference of its orbit = one year = $2\pi R/v$ which yields t = (one year) (v/c) = (one year)/10,000 =16 min 40 sec.

Bradley's discovery finished the revolution that began a half-century earlier.[1] During that time, all of Europe learned that light has an incredibly fast speed, which actually can be measured. The notion that light traveled at a finite speed evolved from a plausible concept based on the eclipses of Jupiter's moons to essentially an article of faith that has never since been shaken.

---

[1] Historical Note: In retrospect, it seems that aberration may have been observed first, but not understood, in careful studies of the location of Polaris, the north star, in the 1650s by one of our heroes from Chapter 4, Jean Picard. Polaris is offset slightly from true north and moves in a small daily circle about true north as the Earth rotates. Picard had sought to observe the parallax of Polaris as a change in size of this daily circle with the seasons, i.e. as the Earth moved about the Sun. Polaris is a good choice to look for parallax (or stellar aberration, for that matter, but no one was thinking of *that* at the time) because deviation from true north could be pinpointed so well, and because refractive bending by the atmosphere would tend to average out during one daily rotation of the Earth. Picard indeed observed deviations, but found them inconsistent with parallax. He observed maximum deviations of about 40 arcsec over a half year, very close to what aberration produces. Of course, at that time, well before Rømer came to Paris, no one had any thoughts about the speed of light. This helps us to see how necessary it probably was to have a concept of the speed of light already in hand from Rømer in order for anyone to come up with the idea of aberration to explain Bradley's data (or Picard's much earlier data). Picard recorded his observations but with little comment.

There was another effect lurking in the Wanstead data, not as revolutionary as stellar aberration, but earth-shaking in its own way. As good as these data match the explanation derived from the aberration (Fig. 8), Bradley noticed a tiny deviation, too small to see in one year with one star, but that revealed itself when the data for all the stars were combined. The stars did not come back to exactly the same north/south position after one year, as predicted by the aberration, but showed a tiny extra displacement. By now, we know Bradley well enough to be certain he would not let this hint go unheeded, but instead follow up on it with his usual ability and steadfast determination. In fact, among his other activities, he continued observing these stars with the Wanstead telescope until 1747, *some 23 years* after he and Molyneux first began their observations. He saw this extra angle continued to grow year after year, to a total of 9 arcsec over a 10-year period before it began decreasing again. By then he was certain this motion was due to *nutation*, a sort of "nodding" of the Earth's rotation axis due to the differential gravitational pull of the Moon on the Earth's oblate form, an effect predicted by Newton.

Bradley's two fundamental discoveries in astronomy, the aberration of light and the nutation of the Earth's axis, were called later, in 1821, "the most brilliant and useful of the century" by Jean Baptiste Joseph Delambre, a historian of astronomy, mathematical astronomer, and director of the Paris Observatory, in his history of astronomy in the 18th century. He said further "It is to these two discoveries by Bradley that we owe the exactness of modern astronomy. .... This double service assures to their discoverer the most distinguished place [among] the greatest astronomers of all ages and all countries." Such a panegyric on the Englishman Bradley by a Frenchman no less! More modern commentators have echoed Delambre's views and noted Bradley, by both his consummate skill as an observer and his insistence on measuring and understanding the experimental environment, set a standard of precision that has inspired experimentalists in the physical sciences ever since.[2]

---

[2] In crediting James Bradley for revolutionary precision in astronomy, we cannot overlook Samuel Molynaux for promoting and enabling the observations at the beginning, and perhaps even more the superb craftsman, George Graham, who built and tested essentially all the instruments Bradley used for his measurements from 1726 onward as well as those used by Astronomer Royal Edmund Halley.

Fig. 10.   James Bradley 1745, at the height of his fame as the 3rd Astronomer Royal of England.

Bradley was appointed the 3rd Astronomer Royal upon the death of Edmund Halley in 1742. Two years later, at age 51, Bradley married and for the rest of his active life resided at Greenwich in the Royal Observatory designed by Christopher Wren. Wren's beautiful Octogan room there still has on display the Wanstead zenith telescope tube used for the published observations of aberration and to prove the Earth's nutation.

At Greenwich, as astronomer royal, Bradley compiled a new catalog of star positions from some 60,000 observations that was published posthumously. F. W. Bessel's catalog in 1818, with 3,000 star positions, was largely based on Bradley's observations. Bradley's health failed, and he retired to Chalford, Gloucestershire, where he died on July 13, 1762.

There is of course one thing missing. Bradley, despite his prodigious abilities and many successes, did not discover stellar parallax. No one else did either, until the next century. And no wonder. The tiny apparent shift of position was hidden among so many larger effects such as refraction of the incoming starlight by the atmosphere and of course stellar aberration itself. Telescopes did advance, and chromatic aberration was practically eliminated in two different ways: (1) Reflecting telescopes pioneered in the 17th century by Isaac Newton became competitive in the 18th century and were basically free of chromatic aberration. (2) In refractive telescopes, compound lenses were developed in which dispersion of one element of the lens canceled that of another element as we will see in Chapter 8. Still, for some time, atmospheric refraction and stellar aberration defeated attempts at seeing parallax. The solution was to look for the differential parallax of two stars situated close to the same direction in the sky. The nearer star would appear to move relative to the other one as the Earth moved in its orbit, while the two stars would share much the same shifts due to refraction and stellar aberration. The first successful measurements of stellar parallax exploited this concept and were made by Friedrich Bessel in 1838 for the star 61 Cygni and a more distant star close to the same line of sight.

So, at last, scientists in the 19th century got the first inkling of how large the universe is. The nearest star to us (and thus the star with the largest parallax), Proxima Centauri, has a parallax of 0.7687±0.0003 arcsec. And people by then could use the light year as a handy measure, so that meant it was commonly understood that Proxima Centauri is about two light years away. The parallax of a star gets progressively smaller as its distance increases, so parallax could not be used to measure the distance to stars further than about 1,000 light years. What could we use to go further? That is the question that is answered in the following special section.

The First Standard Candle: The Work of Henrietta Leavitt

To go further than parallax can take us, we need to find stars whose *intrinsic* brightness we know; then we can calculate their distance by their *apparent* brightness, since they appear dimmer the further away they are. Distant stars whose intrinsic brightness is known have come to be called **standard candles**. The first standard candle was discovered early in the 20th century by Henrietta Leavitt, an astronomer at the Harvard College Observatory. Leavitt was educated at Oberlin College for one year and then at what

(Continued)

became Radcliffe College, where she received her Bachelor's degree. Her senior year she "discovered" astronomy, but then suffered a disease that left her almost totally deaf. She nevertheless obtained a position at Harvard Observatory and was given a job as one of the "women computers" cataloging the brightness of stars for the equivalent today of about $8 an hour. However, she went far beyond what was expected of her, and is now recognized as a pioneer woman astronomer. Leavitt observed that the intrinsic brightness of a certain type of variable star called a Cepheid variable is firmly related to the period of its variation; the brighter the star, the longer the time interval between when it is dimmest and when it is brightest.[3] In 1910, she pointed out that once the distance to a nearby Cepheid was determined by its parallax, then its intrinsic brightness would be known and the brightness–period relationship would become an intrinsic brightness–period relationship. Then the distance to any Cepheid could be determined from its period and its apparent brightness. The following year Ejnar Hertzsprung indeed used their parallax to measure the distance to several nearby Cepheids. Suddenly, as Henrietta Leavitt predicted, we had a way of measuring the distance of stars out to about 30 million light years, the maximum distance at which the Cepheids could be measured with accuracy. *It is hard to overstate the importance of her discovery.* It was the beginning of astronomers' realization of how immense the universe is. Many galaxies fall within this range, so astronomers could tell their distance by the Cepheids they contain. And the stage was set for a different kind of discovery, the expansion of the universe, correlating the distance to Cepheids with shifts in their *spectral lines*, as described in Chapter 8.

Let us return now to our original subject, the speed of light. We have seen how this speed was first measured by telescopes and astronomy. It at last became possible to design terrestrial experiments that could detect the

---

[3] Leavitt studied the Cepheids in the Megallenic Clouds. She computed her brightness–period relationship for those in the small Megallenic Cloud for which she could make the good approximation that they were all at the same distance and thus their apparent brightness was offset from their intrinsic brightness by the same factor.

Fig. 11.   Henrietta Leavitt (1868–1921). She discovered that Cepheid variable stars have a definite relationship between their period of variation and their luminosity or brightness, making it possible to determine the distance of an individual Cepheid by measuring its period and its apparent brightness. This was a revolutionary breakthrough for astronomy. The Cepheids became the first *standard candles* for measuring the size of our galaxy and the distances to other galaxies. Henrietta Leavitt did not seek publicity; in fact, when the committee began to consider her for the Nobel Prize in 1924, they did not know that she had already died of cancer three years earlier.

speed directly, and by 1830, the speed of light was measured by time of flight over great distances made longer by back and forth reflection. But the astronomy measurements stood alone for a century and a half. They had forever changed the perception of speed. In the next chapter, we will return to the question of what is it that is traveling at such unimaginable speed: waves or particles? Newton had said particles, Huygens had said waves, but now a new argument will be introduced by Thomas Young and Augustin Fresnel that will settle the question, at least until the 20th century.

Chapter 6

1800: Thomas Young Revives the Wave Theory

"The Last Man Who Knew Everything."

The title of a biography of Thomas Young by Andrew Robinson.

By the year 1800, the wave theory of light seemed almost forgotten, despite the great success of Huygens in using it 120 years earlier to explain Snell's Law of refraction. Instead, the rival particle theory of light was accepted with little question. The particle theory had its own explanation of Snell's Law, and above all it was Isaac Newton's Theory. Newton's stature had grown to near god-like status in the 80 years since his death. However, the behavior of light was full of unexplained riddles that pointed to a great secret that so far had not been discovered. For example, a most intriguing riddle was how to explain the brilliant colors associated with thin films that we discussed in Chapter 3, especially Newton's Rings that seemed to show how all these phenomena could be related to the thickness of the film producing them; yet a good explanation had eluded people for over 100 years.

Then Thomas Young discovered the secret in 1801, and it was based on the wave theory. The secret was a property of waves that Young called *wave interference*, a term that has been used ever since, and refers to how two overlapping waves can reinforce or cancel each other. Much of this chapter will be devoted to explaining wave interference and applying it to light. Not only did interference enable Young to solve at last the riddle of Newton's rings and all related effects, but in doing so it enabled him to verify that each of the prism colors is associated with its own specific frequency of a periodic light wave, just as each pitch of sound already was known to correspond to a specific frequency of a periodic sound wave. The wave theory also provided a basis for understanding the discovery by Sir William Hershel and others in 1800 that some invisible radiation exists extending beyond the visible spectrum from the Sun, called the infrared (IR) and ultraviolet (UV) beyond the red and violet, respectively. In time it was recognized, as we will see, that the IR is composed of longer wavelengths than the visible and the UV of shorter wavelengths.

Young's work began what might be called the light wave revolution. It was not generally accepted, however, until the crucial experiments and theory of Augustin Fresnel, who arrived on the scene shortly after Young and whose work we will cover in the next chapter. With Young and Fresnel together, the light wave revolution was complete in just 25 years.

In this chapter, we start with the first part of the story, the work of Thomas Young. His ingenious, revolutionary idea of interference began it all. What about the rest of Young's resume? It was impressive. He originated the three-receptor model of human color vision (the RGB values for our printing with computers), he devised Young's modulus, the basic elastic quantity tabulated for all solids, and (well known beyond the world of physical science) he was the first to decipher a significant part of the Rosetta Stone.[1] He was also an excellent dancer. Indeed, he was a Renaissance Man. As noted in the epigraph above, a 2004 book about Young had the subtitle: *The Last Man Who Knew Everything.*

Thomas Young was born in England on June 16, 1773, near Taunton into a devout Quaker family. He was a remarkably precocious child and

---

[1]The Rosetta Stone contains the same decree in three versions. Two are in ancient Egyptian, hieroglyph and demotic script, respectively, and the 3rd in ancient Greek allowing the ancient Egyptian to be translated in terms of the known ancient Greek. Young started his Egyptology work in 1813. He deciphered the demotic script, identifying it as being composed of both ideographic and phonetic signs, correctly found the value of six hieroglyphic signs, but did not deduce the grammar of the language. This early work was crucial to the Frenchman, Jean-François Champollion, who went much further and in 1822 published a translation of the hieroglyphs and the key to the grammatical system. Champollion's decipherment has been the basis for all further developments in the field, and he is regarded as the "Founder and Father of Egyptology".

Fig. 1. Thomas Young.

by age 13 showed a working knowledge of Greek, Latin, French, Italian, and Hebrew, as well as physics, plus competence in the use of the lathe and the construction of optical instruments such as telescopes. Then, after five years with a private tutor he had improved his command of these languages and his knowledge in all scientific fields. He decided on medicine as his career, and came to London to study it at the suggestion of an uncle Richard Brocklesby M.D. who practiced there and introduced Young to a remarkable circle of his friends, including Edmund Burke, Samuel Johnson, James Boswell, and Joshua Reynolds. During Young's two years in London he published a paper on the eye, in which he gave cogent arguments in favor of the theory that explained accommodation of the eye to distance by the contraction of the eyeball. This paper led to his election to the Royal Society. He then continued his medical studies at Edinburgh and Göttingen in Germany, while gradually broadening his activities beyond the strict regulations of his Quaker upbringing. He attended plays and wrote his mother: "I know you are determined to discourage my dancing and singing, and I am determined to pay no regard whatever to what you say... as to dancing the die is cast." Finally, he attended Cambridge, completing his medical degree and entering into practice in London. In London he was only moderately successful in attracting patients, but he was hugely successful in his study of wave motion and in his revolutionary development of the wave theory of light.

Young's understanding of ripples on the surface of water and sound waves in air led to his concept of wave interference, and thence to his ideas about the wave nature of light. Like Huygens before him he thought of a light wave as a disturbance moving in the aether, similar to a sound wave in air, and it became the prevailing view after Young for much of the 19th century. This view will suffice for us for the time being as well, until we learn that light is an electromagnetic wave in Chapter 9. In fact, everything that we say about the interference of light here remains true in the electromagnetic picture as well. As we did in Chapter 2, once again we will stress that the *disturbance* of the aether is moving with the speed of the wave, not the aether itself. Figure 2(a) shows a sketch of a periodic wave representing a moving light wave in the aether, much as a moving ripple on the surface of a stationary pond. Note that light is pictured as a *transverse* vibration of the ether, that is a vibration perpendicular to the direction the wave is moving.[2] The wavelength λ is the distance between two adjacent crests (or equivalently, between adjacent troughs), the frequency f is the number of crests that propagate past any point per second, and the amplitude is the height of a crest (equal to the depth of a trough). Young knew that sound of a single pitch produces a wave in air of a single wavelength and he made the assumption that light of a single color is a wave of a single wavelength as well, and hence light of many colors is composed of many wavelengths.[3]

Recall that a great problem that had been left unsolved for over a hundred years was an explanation of the vivid colors in light reflected from thin films as we saw in Chapter 3. Newton had produced his rings showing how this effect depended upon the film thickness, but the reason for all this remained a mystery. Newton did offer a reason based on his theory of fits, but neither he nor his followers over the next 100 years had

---

[2] Young (and Fresnel) initially thought of a light wave as being a longitudinal vibration in the ether just like a sound wave is a longitudinal vibration in air. For consistency we will picture light as a transverse wave from the beginning, and describe how it became recognized as the correct view in Chapter 7.

[3] Newton briefly entertained this thought about light over 100 years earlier, before he adopted the particle theory, and Young pointed this out in his early publications on light.

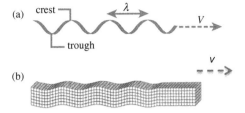

Fig. 2. (a) A simplified representation showing one line of a periodic transverse wave propagating at speed v. A crest and a trough are labeled. The wavelength λ is the distance between adjacent crests (or equivalently between any point and the next point of the same phase). The frequency f is the number of crests that pass a point per second, or $v = \lambda f$. (b) A three-dimensional (3D) picture of the simplified wave in (a).

Note: For simplicity we will usually represent the wave by the transverse displacement of a line of the medium along the direction of propagation of the wave like a wave on a string. However, the medium actually extends above, below, and to the side of the wave shown; it is 3-dimensional. Nearby lines of the medium have about the same pattern for all the waves we will consider, so the medium looks more like Fig. 2(b). We will consider the 3D properties of waves in Chapter 7.

made any progress with fits. Then at last came Thomas Young in 1801 with his entirely new idea of wave interference, illustrated very well in the reflection from a soap film that is illuminated by light of a single color (i.e., a single wavelength) as shown in Fig. 3(a). (Young carried out this experiment, selecting single colors of sunlight with a prism.) The reflected light from the film consists of two beams of about equal intensity, one reflected from the front face and one from the rear face of the film. Where these beams overlap strange things can happen, as can be seen in detail in Fig. 3(b). If they are in-phase with each other, i.e. the crests of one overlap with the crests of the other (and thus the troughs overlap, too), they reinforce each other and the total reflection is bright. Young called this *constructive* interference. On the other hand, if they are out of phase with each other, i.e. crests of one beam overlap with troughs of the other, then they tend to cancel each other out, and there is little or no reflected light. Light canceling light! Young called this *destructive* interference.

In Young's version of the experiment of Fig. 3(a), a soap film was produced across the top of a wine-glass by dipping the glass upside down in soapy water and then tipping it so that the film was vertical; this allowed the film to drain, making it thinner at the top than at the bottom. In a snap shot of a similar experiment in yellow light shown in Fig. 3(c),

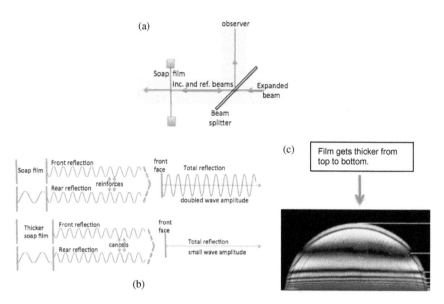

Fig. 3. (a) An experiment with a light source illuminating a soap film through a beam splitter that enables the reflected beams to be split off from the incident beam and seen by an observer. (b) Shows how the front and rear reflections from the soap film combine when they are in-phase with each other producing constructive interference and when they are out of phase with each other producing destructive interference and canceling each other out. (c) What the reflected beam in the experiment in (a) looks like. See text for details.

the front and rear reflections happen to be in-phase at the top, causing the total reflected light to be bright; the interference is constructive. As we look lower, the total reflection goes dark where the back reflection has traveled an additional half-wavelength further than the front one because the film is thicker, canceling the front reflection and producing destructive interference. It is instructive to compare the sketches in Fig. 3(b) with the actual experimental results in Fig. 3(c). Further down the film, the back reflection has traveled another half-wavelength further than the front, putting the two waves back in phase and making the reflection bright again, and this alternation is repeated many times. In a separate experiment, when Young observed all the colors of sunlight reflected from the film, the different colors each had a different wavelength and he saw bands of color where individual wavelengths interfered constructively, much as was depicted in Chapter 3, Fig. 3(a). All these results had been known for over a century, but not with the wave interpretation — until Young.

So, if light is a wave, two overlapping waves of the same wavelength can reinforce each other or destroy each other, or can partially reinforce or partially cancel each other, depending upon their relative phase and amplitude. This is what Young called wave interference. And he saw it with the soap film. And now we must add an important additional point. *The two light beams must come from the same source.* The two beams reflected above actually come ultimately from the same source; *then* one is reflected from the front, the other from the rear surface of the soap film. The two beams are said then to be *coherent* with each other. They have identically the same wavelength and fixed relative phase. In all the examples of interference we show later (and there will be many in this and future chapters) interfering beams originate in the same source.

Young surmised that, just as in the soap film, the patterns seen in the light reflected from thin films of many kinds could be explained by interference. To check this idea quantitatively, however, Young needed to measure the thickness of the film at each point, which is usually not easy to do, but Newton had long before solved the problem of obtaining the thickness of a similar film, the "film" or spacing of air in Newton's rings. This was an experiment described in Chapter 3 and depicted in Fig. 5 of that chapter for incident white light. It is shown again here in Fig. 4 for incident yellow light illuminating the small space between the curved lens and the flat glass plate. (Note that unlike Fig. 3 above, the film is now horizontal and the light beams are vertical.) We summarize the earlier discussion of Newton's rings, but now with a wave interpretation. The total wave of interest is made up of two reflected waves, the wave reflected from the flat surface of the air film and the one reflected from the curved surface of the air film. These two waves overlap and the net wave will depend upon the relative phase of the two waves, just as it does in the soap film.

As shown in Fig. 4, because of the curved surface of the lens, the bottom wave travels a greater and greater distance the further from the center it occurs, changing its phase relative to the top wave and making the two waves alternate between constructive and destructive interference. This is *exactly* what is observed in Newton's rings. The bottom and top reflections produce alternating light and dark regions depending upon the distance from the center i.e. the ring radius. But unlike the changes in thickness for the soap film of Fig. 3(c), the gap thickness now as a function of the radius is known! It is determined by the lens curvature and is

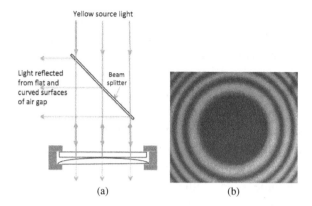

(a)   (b)

Fig. 4. (a) Newton's Rings set up with a single (yellow) color source of incident light. (b) Reflection pattern, which consists of alternating yellow and dark rings. Contrast with Fig. 5 in Ch. 3.

Source: Optics and Lasers in Engineering, Volume 74, November 2015, pg 1–16.

thus readily measured; this was the genius of Newton's experiment. He determined the gap thicknesses for each ring of a given color which told Young over 100 years later the value of λ for that color!

Let us elaborate a bit more on this amazing thing; that Young could measure the wavelength of a color from the diameter of its rings! Suppose we pick a bright ring, which means that the bottom and top waves are in-phase there. We should get another bright ring if we look further out where the extra distance that the bottom wave has traveled (i.e. twice the gap thickness) has increased by a complete wavelength λ. Since the change in gap thickness between the positions of the two adjacent rings was known, it told Young the value of λ. Young found that each color of light was associated with a definite value of λ, and the value of λ increased continuously as the prismatic color changed from blue to red. So, in this and other experiments, Young verified that wavelength is connected with color in the case of light, just as wavelength is connected with pitch in the case of sound in air, though in light the wavelength is much shorter.[4]

Young found that the length of a wave at the extreme red of the prismatic colors is equal to 665 nm (1 nm = 1 nanometer = 1 billionth of a

---

[4]The pitch of sound is determined by its frequency *f*. For instance, middle A above middle C has a frequency of 440 Hz (1 Hz is 1 cycle per sec). In a sound wave in air, *f* is the number of wavelengths that pass a point per sec.

meter) and at the extreme violet end is 420 nm.[5] So light has very short wavelengths, the shortest by far of anything that had been measured to that time. Newton's Rings gave the most accurate measurement of the wavelength of light for decades, and even in 1854, almost 200 years after the measurements were performed by Newton and 50 years after they were correctly interpreted by Young, they are quoted by Mary Somerville[6] in her widely used textbook together with the following comment:

"The determination of these minute portions of space, which have a real existence, being the actual results of measurement, do as much honor to the genius of Newton as that of the law of gravitation."[7]

These wavelengths determined by Young were of course for visible light. At about the same time, Sir William Hershel made a far-reaching discovery studying the spectrum of sunlight spread out by a prism. He placed the bulb of a thermometer at various locations in the spectrum and found that it read an elevated temperature at all locations. To Hershel's surprise, the thermometer read an even higher temperature when it was moved past the red end of the spectrum *where no light could be seen.* He had discovered infrared (IR) radiation, invisible to the eye, but contained in the sunbeam. In subsequent experiments with a glass lens he found that this IR beam, detected by the thermometer, obeyed Snell's Laws of reflection and refraction, but with an index of refraction in glass that was less than in the visible. Eventually, it was realized after the wave theory of

---

[5] Actually, it is the frequency of a light wave $f = c/\lambda$ that determines its color just as the frequency of sound determines the pitch. The wavelength and speed of a given color depends upon the index of refraction, n, of the material the wave is in, but the frequency of the color is always the same. In Newton's rings we measure the wavelength in the air space between the plate and the lens, which is almost the same as the free space wavelength, since n (air) ~ 1.

[6] Mary Somerville was an outstanding physical scientist. She and astronomer Caroline Hershel were the first two women to be elected honorary members of the Royal Astronomical Society.

[7] Notice how Mary Somerville has had no trouble in turning Young's *wave* explanation of the rings into a paean to the achievement of Newton, champion of the *particle* theory of light. This is possible because Newton's Rings were so accurate, a truly great experiment by a great experimentalist even if done over 150 years earlier with a particle theory of light in mind.

light was accepted that this IR beam was simply light of longer wavelength, and had different properties than visible light because of this wavelength. A similar thing happened when J. W. Ritter discovered that the action of light on silver chloride extended beyond the violet, revealing the existence of ultraviolet (UV) light that had shorter wavelengths than visible light which, like IR light, is invisible to the naked eye.

So let us ask, were Young's successes with the wave theory and interference embraced by his peers? Mary Somerville praised the application to Newton's Rings but hers was a comment made years later, in the 1850s. Instead, in the immediate aftermath of his explanation of all these thin film phenomena by wave interference there was no triumphant acceptance of Young's wave theory by the scientific community. Why? To begin with, there was the towering figure of Newton whose followers, initially the vast majority of scientists, had accepted his explanations of these same phenomena using the particle theory, imperfect though they were. Mainly though it just would take some time to understand and accept the totally new explanations by the wave theory.

There was, however, one quite hostile reaction that for a while was more influential than it deserved to be. In 1802, Henry Brougham, an aspiring young scientist and later Lord Chancellor of England, launched a very public attack on Young's work. Brougham had a grudge prompted by an essay Young published in 1800 in *British Magazine* entitled "An Essay on Cycloidal Curves." This essay would have elicited little note, except for three references Young made to a paper Brougham printed in the Philosophical Transactions in 1798. In the paper, Brougham had presented a number of solutions to geometric problems without providing figures or diagrams, and Young was condescendingly critical of his work:

> "We see an author exerting all his ingenuity in order to avoid every idea that has the least tincture of geometry, when he obliges us to toil through immense volumes filled with all manner of literal characters, without a single diagram to diversify the prospect. We may observe with the less surprise that such an author appears to be confined in his conception of the most elementary doctrines, and that he fancies he has made an improvement of consequence, when, in fact, he is only viewing an old subject in a new disguise."

After this criticism of his work, Brougham wanted revenge and he had a bent for invective that would serve him well in his political career. As one of the founders of the *Edinburgh Review*, he had the opportunity to publicly repay Young for the criticisms. He did so in three separate articles, one for each optical essay Young published in the *Philosophical Transactions*. As an emission theorist, that is, a believer in Newton's particle theory of light, Brougham could be expected to criticize Young's findings and hypotheses, but as an anonymous reviewer, he took his criticisms into the realm of invective-filled personal attacks, of which the following is but a very small sample:

> "It is difficult to argue with an author whose mind is filled with a medium of so fickle and vibratory a nature. Were we to take the trouble of refuting him, he might tell us, 'My opinion is changed, and I have abandoned that hypothesis; but here is another for you.' Is the world of science, which Newton once illuminated, to be as changeable in its modes, as the world of taste directed by a pampered fop? ...... In our Second Number, we exposed the absurdity of this writer's 'Law of interference,' as it pleases him to call one of the most incomprehensible suppositions that we remember to have met with in the history of human hypotheses."

Although Young wrote a rebuttal, and certainly we recognize Brougham was way off the mark in his criticism of Young's principle of interference, still the public ridicule by Brougham seemed to stick more than it deserved, in part because there was Newton's authority and Brougham aligned himself with Newton's particle theory of light. Yet, compared with Young's elegant wave solution to the phenomena we have described, including Newton's rings, there was in the particle theory only Newton's theory of fits[8] described briefly in Chapter 3 that was still being worked on by loyal followers of Newton such as Brougham, with no demonstrable success in explaining the same phenomena after nearly a century.

---

[8] It is remarkable that Henry Brougham despite his very successful political career was still finding time to work, albeit unsuccessfully, on the theory of fits to prove the particle theory, even in the 1850s long after the wave theory had been accepted by mainstream scientists.

Brougham was an extreme case, but actually no major scientist converted to Young's wave theory for many years; they found it interesting but not convincing. During that time, in 1807, Young conceived an experiment that he felt would surely convert everyone — the interference of light passing through two slits.[9] It has become THE classic experiment illustrating the wave properties of light, and is now universally called *Young's Experiment*; but in fact it is almost certain *Young never succeeded in doing it*.[10] It was done first by Fresnel and we discuss it in the next chapter where we will see it offers perhaps the clearest example of optical interference.

Young had started an experiment somewhat similar to the double-slit experiment but several years earlier, in 1802, when he investigated the reflection of light by parallel grooves etched in a glass surface. Unfortunately, he didn't make the progress he might have on this method so he didn't recognize the potential of what would become one of the most powerful techniques for measuring wavelengths of light, one that is still useful today. It is now called a *diffraction grating* or *Grating Spectrometer* and is described in Chapter 8.

One thing Young got conceptually wrong was diffraction; that is, the behavior of light when it passes through small holes or encounters small obstacles. For example, he knew that a beam of light expands after it passes

---

[9]The genesis of this experiment was an earlier one by Young in which he placed a "slip of card" edge-on in a narrow light beam, splitting the beam into two beams, one on each side of the card. Young reported seeing a pattern which he attributed to interference between the two beams. He thought, however, that the interpretation of this experiment was open to some criticism, hence his initial delight in conceiving of the double-slit experiment. The slip of card experiment is similar to one by Fresnel on the diffraction of a single wire explained in Chapter 7.

[10]That is the view of most current historians of science. For example, John Worrall bases his negative view on the following undoubted facts: Young does not explicitly state that he did the experiment; Young provides no numerical data; Young says nothing about the light source he uses and the other experimental conditions; and Young never again refers to the experiment. His observations, such as they were, are buried in the middle of a huge mass of other work. Another historian, Nehum Kipnis, believes Young mistook the single-slit diffraction pattern that we will discuss in the next chapter for the two-slit interference pattern, but was dismayed when the pattern was independent of the separation of the slits and gave up work on it. For further discussion and references, see Kipnis, *History of the Principle of Interference of Light*, Birkhauser-Verlag (1991) pp 118–124.

through a very small hole, but he thought this was a result of the light bouncing off the edges of the hole or being attracted to the edges by some mysterious force, the same as the particle theorists thought. As was mentioned at the beginning of this chapter, it was left to Fresnel 10 years later to figure out that most such spreading is an inherent attribute of wave motion. We will see in the next chapter how Fresnel developed a complete theory of light, including diffraction, and successfully applied his theory to the behavior of light in all circumstances. It was only then that the wave theory of light began to be generally accepted. So we will skip most of Young's work on diffraction, in favor of presenting Fresnel's complete theory in Chapter 7.

Let us conclude this chapter by turning back to another of Young's triumphs. Nature has provided us with an example of optical interference on a large scale, an example that Young was the first to understand. It was one of the most impressive of Young's successes, his ingenious use of interference to explain *supernumerary* rainbows, briefly published in 1803 and somewhat more fully explained in 1817, although it is doubtful that many (besides Fresnel) understood either publication while Young was still alive. Many readers have probably seen these rare but glorious faint rainbows which occur beneath the primary rainbow as in Fig. 5, or even more rarely above the secondary rainbow. Once you have seen an interference pattern, the supernumerary rainbow looks like one, on a very large scale — and it is. Just as with the primary rainbow the action takes place in individual drops, but now the drops must be close to the same size in a region of the sky for the same pattern in each of them to be seen. Young proposed the correct explanation which we briefly summarize here for the technically disposed.

From the description of the rainbow in Chapter 1, the reader may recall the *Descartes Ray* in each raindrop, which has the maximum deflection, and the rays near it create the primary rainbow. As we move inward from the primary rainbow at 42.5°, there still are deflected solar rays. Each of these rays at a given deflection angle has a companion ray that emerges at the same deflection but has a different angle of incidence on the drop and follows a different path inside the drop, as shown in Fig. 5, part of which is taken from Fig. 6 of Chapter 1. The two rays have different path lengths in the drop and

(*Continued*)

(Continued)

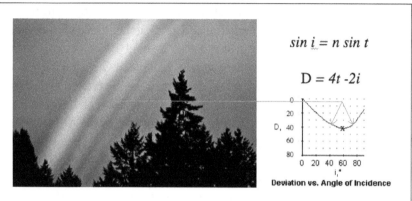

$$sin \underset{\sim}{i} = n \sin t$$

$$D = 4t - 2i$$

Deviation vs. Angle of Incidence

Fig. 5. Supernumerary rainbows plus an explanation showing the two rays at each deviation angle which interfere with each other. At an interference maximum they form one of the supernumerary rainbows, each color or wavelength creating a maximum at a separate angle. *The separation of the colors in the primary is due to dispersion. The separation in each supernumerary is due to interference.* © Larry Andreasen.

hence their wave crests are not necessarily aligned when they leave the drop. In other words, the two rays are light waves that interfere. For some deviation angles the two waves will reinforce each other, for other angles they will cancel, and thus they form an interference pattern. At each deviation angle where the interference is constructive, a supernumerary rainbow is formed. Actually, since the colors in each such rainbow have different wavelengths, they have maxima at different angles spreading the colors in each supernumerary. The figure shows several supernumerary rainbows. Such rainbows are rare, but when the conditions are uniform, producing similar drops over a wide area, there they are; the drops have close to the same pattern and reinforce each other. It is remarkable that Young grasped the essence of the explanation so early, and unfortunate that it wasn't picked up by others for so long.

Before moving on to Fresnel, let us pause and admire Young once more for what he did. It is perhaps hard for us today to imagine how difficult it was to advocate a theory fundamentally at odds with the ideas of Newton who was such an admired and seemingly infallible legend. But

Young did so, and in the case of thin film reflection, convincingly so to the generation following him. His wave explanation of Newton's rings and with it the determination of the wavelengths of visible light became quite compelling as we saw in the popular book by Mary Somerville in 1854. He also ingeniously explained supernumerary rainbows as a product of interference. He was a good experimentalist but not a great one such as Newton or Bradley. In fact, he was at his best when applying his revolutionary ideas to explaining great experiments of others as in Newton's rings or explaining Nature itself as in the supernumerary rainbows. And above all, he introduced this concept of interference, the centerpiece of the wave theory of light.[11] Although Young did not receive the accolades he thought he deserved in his lifetime, posterity has been kind to him, and almost universally today, as we have noted already, the double-slit experiment — even though he almost certainly did not succeed in doing it — is called *Young's Experiment*.

Now it is about time we introduce Fresnel!

---

[11] Interference, physicist Richard Feynman has said, is "the only thing one really needs to understand about quantum mechanics." What? We've only reached 1816 and we're already talking about quantum mechanics? Well, no — Feynman made his statement when he was talking about the *interpretation* of interference by quantum mechanics, which we will discuss in the last chapter.

Chapter 7

The Wave Theory Triumphs: Fresnel 1815–1825

"[I lack] that sensibility, or that vanity, which people call love of glory .... All the compliments that I have received from Arago, Laplace and Biot never gave me so much pleasure as the discovery of a theoretic truth, or the confirmation of a calculation by experiment."

Augustin-Jean Fresnel in a letter to Thomas Young (1824).

Fresnel's accomplishments rank among the greatest achievements in physics in the 19th century. In the last chapter, we saw how Thomas Young laid the basis for the wave theory of light with his concept of interference. Fresnel built a complete theory on this foundation. Physicists often invoke the image of the Holy Grail for the great breakthrough vision they are seeking at that moment. In many ways what Fresnel achieved was just that vision for light. Fresnel's picture of light was the first to account for everything that was known up to that time, but was also a very practical theory that for many decades led to great advances in useful optical instruments. Of course, the vision of light that emerged for Fresnel would eventually be seen to be incomplete, and a more complete vision would be found that included electromagnetic waves and the idea of photons. Those are subjects for later chapters. This chapter is about Fresnel and his vision.

Fresnel was plagued from childhood by tuberculosis, eventually the cause of his early death, and his entire life was a struggle against bodily suffering and fatigue. Nevertheless, he threw himself into difficult tasks, and was armed with the mental strength and character to prevail in his scientific quest. He started his scientific career with the desire to make a difference, and this he certainly did. He was a great experimenter, imaginative in thinking up experiments and gifted in doing them with his hands. Fresnel was also a great theorist who developed the first complete mathematical description of light waves, called the Huygens–Fresnel theory. That theory could answer every question about light that people could think of at the time, including an explanation of *diffraction*, the spreading of light when it passes through narrow holes or encounters small obstacles. Finally, Fresnel (with help from Young) figured out the *polarization* of light waves, namely that the vibration of the wave is perpendicular to the direction in which it propagates. This was the coup de grace that converted any remaining skeptics to the wave theory. In addition to all this, Fresnel possessed a remarkable practical ability to develop useful optical devices, the most famous being Fresnel lenses, which increased the range of lighthouse beams dramatically and were quickly deployed worldwide for safer shipping.

Augustin-Jean Fresnel was born in 1788 at Broglie in Normandy, the son of Jacques Fresnel and Augustine Mérimée. Jacques was an architect then undertaking major improvements on the chateaux Victor-François de Broglie, the Second Duke and marshal of France under Louis XV and Louis XVI.[1] When Jacques completed the chateaux project, the family moved to Cherbourg on the coast where Jacques worked on construction of the harbor until the work was suspended due to the Reign of Terror in 1794 and the family moved to another spot in Normandy, Mathieu, near Caen. Augustin's parents were devout members of the Roman Catholic reform movement called Jansenists, and Augustin was brought up in a stern atmosphere with strict Jansenist values, which placed the highest merit on personal achievement, performance of duty, and service to society. Serious and intent, Fresnel bound himself closely to these ideals, shunning pleasures and working to the point of exhaustion. Although

---

[1] The author cannot resist pointing out the coincidence that Augustin Fresnel, who developed the complete theory of the wave nature of light, was born on de Broglie's estate about 100 years before the birth there of Louis de Broglie who first conceived of the wave nature of material particles such as electrons. We will return to de Broglie in Chapter 10.

Tuberculosis cast a shadow over Fresnel's entire career, lending urgency to everything he attempted, Fresnel was always attentive to detail, systematic, and thorough. In science — and in all walks of life — he held tenaciously to his convictions and defended them with courage and vigor.

His early education was under his parents, but he still could not read when he was eight years old. Nevertheless, at 13 he entered the École Centrale in Caen, where he began to excel, particularly in science and mathematics. He decided on an engineering career, and at 16 entered the École Polytechnique in Paris, where he acquitted himself with distinction. From there he went to the École des Ponts et Chaussées and graduated as a civil engineer in the year Napoleon became emperor. He was assigned to road building in the Vendee region, on the Atlantic coast, and in 1812 to a major road connecting Spain with northern Italy through France. Still, he began scientific work in his spare time, particularly on the subject which fascinated him the most — light. Meanwhile, Napoleon had been exiled on Elba, but when he escaped and landed at Cannes, Fresnel was so upset that he left his job and set off on horseback, offering to fight for the King against Napoleon. But when Napoleon regained control of France, Fresnel lost his engineering post and was put under police surveillance. His only option was to return to his parents' home at Mathieu. On the positive side, Fresnel suddenly had free time for his work on light and even to carry out some crucial experiments which convinced him of the wave nature of light.

Soon Napoleon was defeated at Waterloo, and Fresnel was reinstated and given an engineering post in Rennes. The important advances in the study of light he had made during his relatively brief time at home Fresnel wrote up and sent to Paris.[2] In this unorthodox way he won Francois Arago, already a famous member of the French Academy of Sciences, to the cause of wave optics. From then on Arago became an influential champion of Fresnel's research and provided him with translations of Young and others (since Fresnel knew only French). Fresnel repeatedly requested leave from his post in Rennes to go to Paris and continue his

---

[2] Much of his initial work on light was undertaken without knowledge of the latest contributions by other scientists. He did not know of the wave theories that had been postulated by Young or even much earlier by Huygens, nor did he know of the latest developments in the corpuscular theory supported by the majority of scientists.

optical investigations, and with Arago's influence his leaves became frequent and extended. Arago stands out in this story for his quick recognition of the importance of Fresnel's theory in its infancy and his influence in enabling and promoting the work.[3]

From almost the beginning of his investigations (about 1814) Fresnel had concluded that light is a wave. He had become convinced when he placed a wire across a sunbeam and inspected the shadow of the wire on a screen: alternate light and dark lines, called fringes, appeared in the shadow, as shown in Fig. 2. The fringes were nothing new; Grimaldi, whom we will talk about later in the chapter, had observed them in the 1600s. What impressed Fresnel was that certain fringes, for instance, the bright, narrow one pointed out by the arrow in Fig. 2, disappeared when he blocked the beam on one side or the other of the wire. In that case, just the light that had leaked into the shadow from the lighted side of the wire could be seen, and the bright fringe was gone. The fringe required light from both sides. In other words, the light from the two sides *interfered* with each other.[4] This observation convinced him that light could only be

Fig. 1.   Portrait of Augustin-Jean Fresnel from the frontispiece of his collected works (1866).

---

[3] Arago's life is remarkable: he was an adventurer, scientist, administrator, and, later, prime minister of France.

[4] In principle this observation was similar to, but without the problems of, Young's earlier "slip of card" experiment alluded to in the previous chapter.

Fig. 2. Shadow of a vertical wire intercepting a single wavelength beam. The arrow points to a feature at the center of the shadow where the light from each side produces an interference maximum bordered by dark interference minima. The central bright feature disappears when light from either side of the wire is blocked.

a wave, as Young had been saying for 12 years but not known to Fresnel. When Fresnel finally learned from Arago about Young's work, and his use of the term *interference* to describe similar effects, he adopted that term, and acknowledged Young's precedence.

Still he proceeded with single-minded determination to carry out other experiments to demonstrate the wave nature of light, which he understood from Arago was not generally accepted. Among these was the *double-slit experiment* already mentioned in the previous chapter now known mistakenly as Young's Experiment. In his *prize memoir* (which we will discuss later) Fresnel summarized this experiment in just a single paragraph as follows:

"Brighter and sharper fringes may be produced by cutting two parallel slits close together in a piece of cardboard or a sheet of metal, and placing the screen thus prepared in front of the luminous point [i.e. a small lamp or a beam of sunlight shining through a small hole]. We may then observe, by use of a magnifying-glass between the (slits of the) opaque body and the eye, that the shadow is filled with a large number of very sharp-colored fringes so long as the light shines through both openings at the same time, but these disappear whenever the light is cut off from one of the slits."

This is disarmingly simple for a difficult experiment that Young himself did not succeed in doing. Fresnel certainly devotes more detail in his

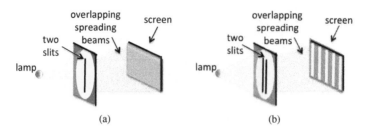

Fig. 3. Young's proposed two-slit experiment, carried out first in 1816 by Fresnel. (a) First, light from a small yellow lamp is sent through a single narrow slit, creating a beam that spreads out and uniformly illuminates a screen. (b) Next the light is sent through *two* slits, spreads out in two overlapping beams and illuminates the screen, and now something <u>is</u> surprising: alternating bands appear — another example of what we have called an *interference pattern* — just from adding the second slit. Where there are now dark bands, the intensity has actually been reduced by adding light; at those places the beams from each slit have largely canceled each other out! This experiment offers convincing proof of the wave theory of light.

published notes. Figure 3 shows a version of this experiment. In the first step (Fig. 3(a)), light of a single color from a small source passes through a narrow slit, spreads out, and illuminates a screen rather uniformly.[5] Next, a second slit, initially covered, is uncovered so that light from the same source now goes through *two* slits, located side by side as shown in Fig. 3(b). The expanding light forms two overlapping beams that illuminate the screen.

Now the surprise. Instead of simply a brighter version of Fig. 3(a), the screen has become filled with alternating bright and dark bands (or fringes as we also call them) as shown in Fig. 3(b). At the dark bands, **adding** the second beam actually **reduces** the light and even can result in **no** light at all because there the two beams cancel each other out! This behavior of light can be explained by interference just as in the reflection from thin films in the last chapter. At the central bright band the two waves from the slits have traveled the same distance so they are in-phase and that is why

---

[5] The spreading of light by a narrow slit was a common observation, explained by Newton as light deflected by the slit edges. Fresnel knew better; it is due to diffraction which we will take up shortly.

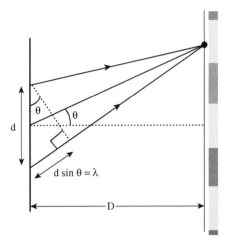

Fig. 4. In the two-slit interference experiment of Fig. 3, the path difference between the beams from the two slits is $d \sin \theta$, which must be an integral number of wavelengths λ to create a maximum on the screen. For the first maximum to the side, this yields the relation $\lambda = d \sin \theta$ as shown. (All of this is true only when $D \gg d$ so that the rays shown are almost parallel to each other.)

it is bright. The adjacent dark bands on each side mark the places where the light from one slit has traveled a half wavelength more than the light from the other slit and produces destructive interference there. Progressing further, one sees alternate bright and dark regions depending upon the difference in distance the two waves have traveled to those regions. This pattern illustrates interference beautifully. A most important point is that the wavelength λ can be measured by determining the angular separation θ between adjacent bright fringes on the screen, and is given by $\lambda = d \sin \theta$ with d the spacing of the slits, as illustrated in Fig. 4. The values of the wavelength for a given color agree with those Young found by Newton's rings, which is very strong support for the wave theory.

We now describe Fresnel's *two-mirror experiment* which demonstrated interference in circumstances where any attractive forces of the slits could not be invoked to explain its effects. Two mirrors arranged end to end at an angle slightly less than 180° were placed in front of a beam of sunlight with a prism used to select a single color as shown in Fig. 5. The reflected beam from each mirror landed on a single screen, where

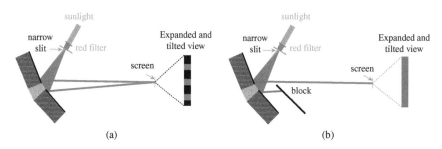

Fig. 5.   The two-mirror experiment showing interference between the beams from two mirrors having a common source.

they overlapped and interfered with each other to produce alternating bright and dark bands as seen in Fig. 5(a). If one of the mirrors was blocked as in Fig. 5(b), however, no bands appeared on the screen, just uniform illumination. The interpretation of the bands in terms of interference is similar to the case of the double-slit experiment. A dark band occurs where a crest of one beam overlaps a trough of the other, a bright band occurs where a crest of one overlaps with a crest of the other. The wavelengths of the different colors found in this way agree exactly with those revealed by the double-slit experiment and by Newton's rings.

These experimental endeavors were only the first of Fresnel's contributions. They clarified and extended Young's concept of interference. What Fresnel did next was far more daunting. He undertook the task of developing a complete theory of light waves. One part of this theory would necessarily be to explain the spreading of light when it goes through a narrow slit or passes around a narrow obstacle. Francesco Maria Grimaldi, an Italian Jesuit priest, mathematician, and physicist from Bologna, had originally observed this spreading of light around a narrow wire in the 17th century, as we mentioned earlier. He called it *diffraction*, but the correct explanation had eluded everyone for over 100 years until Fresnel showed diffraction is a natural consequence of the wave nature of light.

We can produce the diffraction of light by a slit very simply by looking at a fairly distant streetlight or traffic light through the slit formed by pressing together two of our fingers. As we squeeze the fingers together making the slit narrower, we note the light smears out perpendicular to the

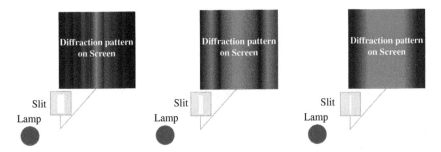

Fig. 6.   Single-slit diffraction patterns. Single wavelength light from a lamp illuminates a screen after passing through a slit. The diffraction patterns are shown for three different slit widths. The central peak gets broader and the sidebands get further apart as the slit is made narrower.

slit, and sidebands appear. As we will learn, this is not the light reflecting off our fingers; rather it is the result of reducing the width of the aperture through which we see the light. This effect is shown more reproducibly in Fig. 6, where slits of varying width are cut carefully in a thin metal sheet through which light — violet in this case — passes and falls on a screen.[6] Fresnel did experiments such as this. For a sufficiently narrow slit, he observed what we see in Fig. 6: the light spreads out on the screen and weak sidebands appear. The narrower the slit, the wider the spreading and the more separated the sidebands.

How did Fresnel explain all this? He began with Huygens' Principle which we introduced in Chapter 2 to describe the propagation of a wave front. The Principle states that a wavefront acts as a source of Huygens' wavelets which create the wavefront further on. Fresnel modified Huygens' Principle so as to describe the propagation of periodic waves, that is, light waves of a specific color (i.e. wavelength) by giving the Huygens wavelets the same wavelength as the wavefront that produces them. (Wavefront now means the points on a wave that have a common phase. For periodic waves, the phase at a given point describes whether

---

[6] It is thought by some scholars that when Thomas Young attempted the double-slit experiment, he actually saw instead this single-slit pattern from each of the slits, which didn't change with the separation between the slits as he was expecting, and this accounts for his frustration and failure to report his results. Good as he was, Young did not understand diffraction.

the wave is at a peak or a valley or at a particular place in between.) The amplitude of the wave at any point beyond the wavefront is the sum of all the wavelets that reach that point from the wavefront, but now the sum is determined *by the relative phases* as well as amplitudes of the wavelets. For example, where two wavelets meet they may have the same amplitude but opposite phase. At that place their sum would be zero. Fresnel made this rough argument very precise, and his formulation of Huygens' Principle became known as the Huygens–Fresnel Principle.[7]

We now apply this idea to the experiment of Fig. 6. In the forward direction, the wavelets from the slit area all travel about equal distances to the screen and hence have about the same phase as one another and therefore reinforce each other. This explains the bright central maximum produced by each slit of Fig. 6. To reach an observation point to the side, however, the wavelets travel different distances depending upon where they originate in the wavefront and thus will have different phases from each other. These wavelets will partly or totally cancel each other out. This explains the reduced amplitudes to the side of the central maximum.[8] These features are summarized in Fig. 7, which gives the single-slit diffraction pattern calculated precisely by Fresnel's theory, showing how the intensity varies with the angle θ of the observation point measured from the central peak. The intensity falls to zero (i.e. the wavelets cancel each other out) when $\sin \theta = \pm \lambda / W$, where λ is the wavelength of the light and W is the width of the slit. (This formula holds only when the screen is so far away that all wavelets from the slit to the same point on the screen may be taken as parallel, the same condition as in Fig. 4.) Note from this formula that the smaller the slit width, the larger the angle and hence the broader the central peak, in agreement with the experiment shown in Fig. 6.

Such diffraction blurs the image produced by a lens at its focus regardless of how good the optical quality of the lens. If the slit above is replaced by a circular lens of diameter *D,* it can be shown that the formula

---

[7] Eventually, Kirchoff modified the principle further to make it conform to the *wave equation* which is above the level of this book.

[8] It is this point that early opponents of the wave theory — even the great Newton — did not consider and therefore thought that if light were a wave it must spread out in all directions when it went through even large holes, which was contrary to observation.

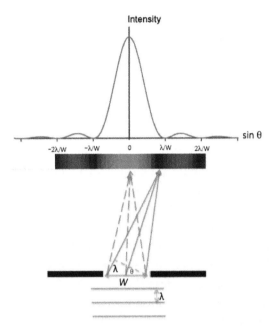

Fig. 7. Plot of intensity vs. sin θ given by Fresnel's theory of single-slit diffraction. All the wavelets from the slit to the center of the screen (represented by dashed lines in the figure) travel almost equal distances, and hence arrive with the same phase and produce a maximum of intensity. The wavelets to the dark part of the screen travel different distances arriving with different phases such that they cancel. Such an analysis reproduces all the features seen on the screen in Fig. 6.

for the width of the central diffraction peak is very similar; it becomes sin $\theta = 1.22\ \lambda/D$. This focal blurring thus gets smaller the larger the diameter of the lens. This is one reason to make a telescope lens large; the Hubble space telescope for example has a diameter of 2.4 m. Of course, another reason is that a large lens collects more light and thus it may allow us to see a faint object even if we can't resolve it.

Now let's turn to another example illustrating diffraction, namely the bending of light around an obstacle, which turned out to be crucial to the acceptance of Fresnel's wave theory. In 1818, the French Academy of Sciences launched a competition to explain the properties of light and at Arago's urging Fresnel, who was still employed as a civil engineer, entered this competition by submitting his new wave theory of light. (This is the memoir we mentioned in connection with the double-slit experiment.) In 1819, the committee to judge the Grand Prix of the Académie des Sciences, with Arago as chairman, and including Poisson, Biot, and Laplace, all eminent scientists, met to consider the submissions. All members of the committee other than Arago believed in the corpuscular model of light. Poisson in particular was noted for his mathematical ability and succeeded in computing some of the implications of Fresnel's theory. He thought he had discovered a flaw when he found that Fresnel's theory predicted that a small light source illuminating an opaque circular screen would produce a small bright spot on the axis *behind* the screen, presumably where "common sense" suggested that the screen should block the light entirely. Poisson interpreted this as an absurd result that disproved the wave theory. But Arago, ever faithful to Fresnel, thought this "flaw" should be tested. With great care he carried out the experiment and found to his own (Arago's) surprise but not to Fresnel's, that the spot exists! It is shown in Fig. 8.

We can give a rough explanation here. First, it is similar to the interference peak seen at the center of the shadow of the wire shown in Fig. 2, but for a circular edge rather than the straight edges of the wire. In the circular case, for the part of the wavefront from the small source that passes unblocked just outside the circular screen, the Huygens–Fresnel Principle says that every point acts as a new point source of light. Those wavelets coming from points near the circumference of the screen and going to the center line of the screen's shadow travel the same distance,

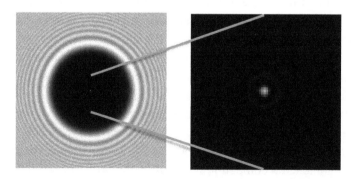

Fig. 8. The spot at the center of the shadow of an opaque circular object, predicted by Fresnel's theory and later observed by Arago.

so all the light passing close by the screen arrives at the center line in-phase and constructively interferes. At all other points in the shadow the wavelets exhibit large cancellation. This results in a bright spot at the shadow's center, where particle theories of light predict that there should be no light at all. However, the edge of the circular screen must be sufficiently smooth, which explains why the bright spot is not usually seen.[9]

Arago's dramatic observation of the spot may have provided the clincher; at any rate Fresnel's memoir won the competition and convinced most scientists that Fresnel's methods were the ones to use when calculating the effects of diffraction. Yet, many of the best known — including Laplace, Biot, and Poisson — remained unconvinced of the wave-nature of light even though that theory provided the only method known for calculating diffraction — and they all used it! What was the stumbling block? It was probably that light waves seemed to require a medium extending over all space to carry the optical vibrations, what we have called the luminiferous aether. This medium revealed itself in no other way and, in particular, offered no measurable resistance to motion through it. What finally convinced almost everyone of the wave theory (despite the aether problem, which would not be solved for many decades) was the success two years later in using the wave theory to understand *polarization* of light.

Polarization refers to the direction of vibration of a wave. If the vibration is along the direction of propagation of the wave, as it is for a sound wave in air, it is called longitudinal. If it is perpendicular to this direction, as it is for a typical surface water wave, it is called transverse. For simplicity we have assumed from the beginning that a light wave is transverse, but Young and Fresnel each initially thought that light is a longitudinal wave, since they imagined the luminiferous aether to be more like a gas than a solid in order to allow material bodies to pass through it with little or no resistance, and a gas cannot support a transverse wave. For example, sound waves in air are longitudinal. A medium has to be a solid to support a transverse wave within it, and they couldn't conceive of a solid aether

---

[9] Arago later noted that the phenomenon (which was to be known as Poisson's Spot or the Spot of Arago) had already been observed — but not explained — by Delisle and Maraldi a century earlier!

that still let any ordinary object move through it undisturbed. That inconvenient point was just ignored when evidence that light is indeed a transverse wave became overwhelming as we will now see. Fresnel in particular came to believe that preconceptions about the supposed aether should not dictate how we formulate the laws of optics; only the phenomena should. And eventually, the idea that light waves are a vibration in a material aether was abandoned in favor of the idea that light is an electromagnetic field as we will discuss in Chapter 9. Even then the aether persisted as the supposed seat of electromagnetic forces until it was found to be unnecessary, as we will see.

A transverse wave is said to be linearly polarized if the vibration is always in one direction. In Fig. 9(a), we represent such a wave by a horizontal line, with an arrow showing its direction of propagation. When the polarization is along an axis parallel to the page, we represent it by vertical arrows and when it is perpendicular to the page, by dots, as in Fig. 9(a). An unpolarized wave is represented with both vertical arrows and dots as shown in Fig. 9(b).

The first clue that light is a transverse wave was double refraction, though the clue was not recognized for a long time. In Chapter 2, we mentioned the discovery of double refraction of light in crystals of Iceland spar (now called calcite) in 1668 by the Danish scientist Rasmus Bartholin. An incident light ray splits into two separate rays in the crystal. One ray obeys the usual rules of refraction and Bartholin called that ray ordinary (O); the other ray violates the rules and he called it extraordinary (E). The two rays created two images of an object when looked at through the crystal, as shown in Fig. 10(a). We pointed out in Chapter 2 that Huygens established that the proportion of the two rays depends upon the angle of incidence relative to a special axis of the crystal known as the optic axis,

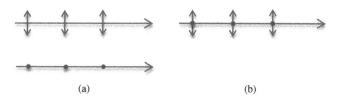

(a)   (b)

Fig. 9. Representations of polarization as described in text.

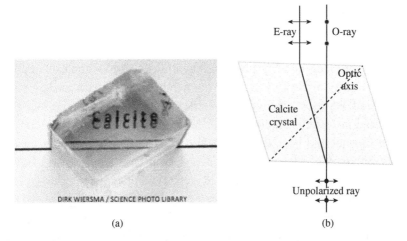

(a)   (b)

Fig. 10. (a) Double refraction shown by a calcite crystal, which is (b) explained if light is assumed transversely polarized with an *O* and an *E* beam, as in the text. The optic axis lies in the plane of the paper perpendicular to the *O* beam polarization.

but there matters stood for over 100 years. Then Fresnel realized that if light is transversely polarized, the direction of polarization of a beam relative to the optic axis could be what made it *O* or *E*. As illustrated in Fig. 10(b), he correctly showed that the polarization of an *O* beam points at right angles to the optic axis, and that of an *E* beam points parallel to the plane containing the optic axis and the beam itself.

From 1816 to 1819, the realization dawned that if light is a totally transverse wave, many other problems could be explained, too. Fresnel had shown by experiments in 1816 that the *O* and *E* beams generated by the same initial light beam, as in Fig. 10(b), do not interfere with each other. In a two-slit interference experiment, for example, when *O* is directed through one slit and *E* through the other, no interference pattern appears, just uniform illumination of the screen. Fresnel explained this observation by pointing out that if *O* and *E* vibrate perpendicular to each other, as they do in Fig. 10(b), they cannot cancel and no interference pattern would be expected. If, however, *O* (or *E*) goes through both slits, then they can cancel and interference will appear, as Fresnel observed. These and other experiments showed that *O* and *E* consist of vibrations that are perpendicular to each other and to the direction of propagation.

Most light we encounter does not seem to have these polarization effects. Yet, it is still a transverse wave. Fresnel came up with the correct view that unpolarized light consists of transverse vibrations, but they are randomly oriented in a light beam emanating from most natural sources, and that is why ordinary beams do not exhibit any polarization.

One of the most striking polarization experiments actually had been performed in 1808 by Étienne–Louis Malus before any of the understanding by Fresnel had been developed. While carrying out precise measurements on double refraction, Malus observed that when a ray of light reflects off many ordinary surfaces at a certain angle, 56° for glass, 53° for water, it behaves like an *O* beam emerging from a calcite crystal, in other words, it is polarized as indicated in Fig. 11. This angle is called the Brewster angle, because Sir David Brewster made careful studies of the phenomenon in 1815. Using the interpretation achieved 10 years later by Fresnel, these results can be expressed as shown in Fig. 11. If the beam is polarized in the plane of incidence, then at the Brewster angle no reflected light appears. If the beam has a component of polarization perpendicular to the plane of incidence, the reflected beam has that direction of polarization. So we have learned of two ways to obtain polarized light from an initially unpolarized beam, by a doubly refracting crystal as in Fig. 10 or by directing the beam at Brewster's angle as in Fig. 11. Yet another way is to use a polarizing sheet that is stretched in such a way that it transmits

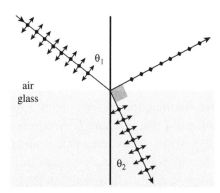

Fig. 11.   Incident light at Brewster's angle produces polarized reflected and transmitted beams.

light polarized along one axis and absorbs light polarized at 90° to that axis. Numerous other ways have been devised for obtaining and studying polarized light.

We will mention only one of many applications of polarized light, but it is one that most people have seen when they use a TV, a computer screen, or a video projector. All screens now use a liquid crystal display (LCD), sometimes backlit by a light-emitting-diode screen. In brief, the operating principle is based on the rotation of linear polarization by an LCD array of small cells (pixels) that is placed between two polarizing sheets oriented 90° with respect to each other. In the natural state, the front sheet blocks the light coming through the back sheet, and the screen is dark. When a voltage is applied to an LCD pixel, it rotates the polarization proportional to the voltage, allowing the front sheet to transmit light at that pixel according to the voltage applied. The total screen (i.e. the LCD array) faithfully creates whatever image is set by the voltage pattern at the pixels.

After 1824, Fresnel devoted less time to his research on the nature of light and more on practical applications. He was employed by the Lighthouse Commission and to this work he brought the same inventiveness, concentration, and perseverance previously manifest in his studies on the nature of light. From this period he is perhaps best known for his invention of the *Fresnel lens*, which is a composite lens of large diameter but short focal length that is made without the thickness required of conventional lenses and is also suitable for 360° illumination from lighthouses. We will not attempt an explanation, but show an example in Fig. 12. This design was adopted in lighthouses, not only in France but all over the world, and has been called "the design that saved a million ships." Fresnel lenses have had countless uses when the lack of a sharp focus (its major drawback) is not an issue. They were employed for decades in automobile headlamps and are currently exploited in a number of applications, from screen projectors to solar energy concentrators.

Fresnel died of tuberculosis in 1827 at the age of 39. He had struggled throughout his life against ill health, but it is remarkable that he was able to undertake an exceptionally high workload despite suffering from severe fatigue. Toward the end of his life he did receive recognition. In 1823, he was elected to the French Académie des Sciences; then he was elected to the Royal Society of London (nominated by his sometime rival Thomas

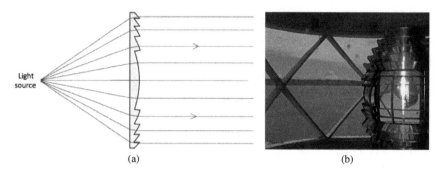

Fig. 12.   (a) Design of a Fresnel lens; (b) a Fresnel lens deployed in lighthouse, Puget Sound, WA. Jet Eliot on WordPress.com.

Young) and he received its Rumford Medal in 1827. He left us a theory of light waves that has been expanded since then but not supplanted. He also laid the foundation for great advances in optical instrumentation in the 19th century, as we will soon see in Chapter 8.

Let us end this chapter by summarizing the wave theory of light as it was developed by Young and Fresnel:

1. A light beam is a transverse wave vibrating in a medium called the aether.
2. A specific frequency of the wave is associated with its own specific color.
3. Two light beams that are split off from the same source show interference when they recombine. If the beams have a single wavelength and the same phase, the interference will be constructive and their amplitudes will add. If they have opposite phases, the interference will be destructive and the amplitudes will cancel.
4. A beam passing through an aperture is diffracted, spreading out in a manner consistent with the rules of interference among the Huygens wavelets from each point of the wavefront at the aperture. The spreading turns out to be larger the smaller the aperture.
5. The speed of light is slower in a material with a larger index of refraction, which explains Snell's Laws. (However, what causes light to move at a different speed inside a material was not discovered until the end of the 19th century, as we will see.)

We will continue to use this theory in Chapter 8. However, important ideas about the nature of light were yet to be discovered, as shown in Chapters 9 and 10.

Chapter 8

Joseph Fraunhofer
Observes Spectral Lines

"In looking at this spectrum for the bright line which I had discovered in
a spectrum of artificial light, I discovered instead an infinite number of
vertical lines of different thicknesses. These lines are darker than the rest
of the spectrum and some of them appear entirely black."

Joseph Fraunhofer, reporting on his revolutionary discovery of
absorption lines in the spectrum of sunlight.

In 1814, in the middle of what we have called the light wave revolution, a
glassmaker in Germany, Joseph Fraunhofer, made a discovery that eventually was to prove the key to understanding how light waves are emitted or
absorbed. He was viewing the dispersion of sunlight into its spectrum of
colors through a glass prism, as Newton and many others had done, but
using the much better glass he had fabricated. He was amazed at what he
saw: superposed on the familiar red through violet spectrum were hundreds of mysterious dark lines never seen before. These lines had always
been there of course, they had just been smeared out by imperfections in
the glass and therefore not resolved.

It is hard to exaggerate the importance of this discovery. Soon flames
and other light sources were found to have lines too, both bright and dark,
and specific lines were eventually associated with emission or absorption

by specific atoms or molecules in the sources. In this chapter, we explore how these *Fraunhofer* lines as they were called, when studied by ever more precise instruments of wave optics, became central to the development of the physical sciences in the 19th century, from the study of stars and galaxies to the structure of atoms. The presence of certain lines in the spectra of stars told astronomers about their chemical composition. The shift in wavelength of the lines in a star revealed its velocity relative to the Earth, and ultimately (in the 20th century) showed that the universe is expanding. On the small scale, the study of the lines of individual elements led to the discovery by the German Johann Balmer in 1885 that all lines of atomic hydrogen could be accounted for by one simple formula. As we will find out in the last chapter, this *Balmer formula*, when combined with the idea of photons, led to the quantum theory of atomic structure. In fact, the mere existence of these lines was a clue, unrecognized of course at the beginning, to the idea of quantum mechanics discovered a hundred years later.

Meanwhile, even the concept of a measurement underwent a radical change. The accuracy of one instrument, the *interferometer* (so named because it is based on the interference of light), was exploited to measure ordinary lengths in terms of the wavelength of a Fraunhofer line. The first such measurement in 1892–1893 was to count the number of wavelengths of the red line of cadmium that fit within the length of the standard meter in Paris. In principle, this means that the length of any object on Earth could be determined by the number of wavelengths of a specified Fraunhofer line it contains, and this length could be communicated anywhere, even to some being on a distant star system, since the Fraunhofer lines presumably would be universal. The combination of interferometers and Fraunhofer lines has been used in countless applications to the present day. We illustrate with the application to the first detection of gravitational waves coming from deep space, which will provide insights in astronomy and astrophysics for years to come.

We begin with the person who started it all, Joseph Fraunhofer (1787–1826), whose life is an inspirational story. He rose from destitution as a child to the pinnacle of his profession of glass-making by his 30s and was the recipient of numerous scientific and academic honors before his early death at age 39. He succeeded because of his brilliant mind, unwavering determination, skillful hands, and more than a little bit of luck. He was the

11th child of Franz Xaver Fraunhofer and Maria Anna Fröhlich, born in Straubing in the Electorate of Bavaria. He descended from glassmakers on both parents' sides. His mother died when he was 10 and his father a year later, so he was orphaned at age 11 and began working as an apprentice to a tyrannical glassmaker, Philipp Anton Weichelsberger in Munich who treated him solely as a laborer. Two years later Weichelsberger's house in which Fraunhofer lived and worked collapsed, gravely injuring Weichelsberger's wife and leaving young Fraunhofer buried in the rubble but miraculously unscathed. Prince-Elector Maximilian Joseph [later Emperor] traveled to the scene to lead the rescue operation and was so impressed by Fraunhofer's survival that he made him a gift of money, which the child saved. Also he ordered his privy councillor to take a personal interest in the boy.

Fraunhofer continued working under Weichelsberger until he was 19, when the latter's rules preventing him from studying optical theory and glass practice at night, and keeping him from attending a school for working class boys, finally led to his buying out the contractual agreement with Weichelsberger using the original gift from the prince elector. The privy councilor brought Fraunhofer into his Optical Institute, a secularized Benedictine monastery devoted to glassmaking — and devoted also to keeping their techniques secret, a long tradition among most glassmakers (a tradition that Fraunhofer maintained). There Fraunhofer learned how to make fine optical glass and began improving on the results until this orphaned boy, whose life seemed so hopeless when he was found in the rubble, in 15 years was making the best lenses and prisms the world had ever seen.

Just at this time there was a huge market in Europe for better optical instruments for surveying. In the early 1800s Napoleon ordered more accurate "topographically and astronomically correct maps" of his newly occupied territories and of new allies such as Bavaria where Fraunhofer lived. These maps served military and civil purposes including an accurate basis for taxation. Fraunhofer became the director of the Optical Institute in 1818, and owing to the fine optical instruments developed by Fraunhofer (and to the blockade in the Napoleonic era as well) Bavaria displaced England as the leader of the optics industry. No one else could rival Fraunhofer.

Fig. 1.   Joseph Fraunhofer (1787–1826).

Producing fine lenses was an ongoing process of eliminating defects and impurities in the glass and of compensating such effects as chromatic aberration more accurately. Chromatic aberration is caused by the dependence of the refractive index upon the color or wavelength, an effect known as dispersion. As we have seen in Chapter 3 and elsewhere, dispersion is what causes a prism to spread white light into colors, but it also results in a lens having a different focal length for different colors, as illustrated in Fig. 2. This was the problem that caused the super-long telescopes to be built in the 1600s as we saw in Chapter 4. By the 1700s people had learned to compensate dispersion using two lenses made of different glasses as shown in Fig. 2. It was in his attempts to make lenses ever freer of chromatic aberration that Fraunhofer made his remarkable discovery.

Fraunhofer had not heard yet of Thomas Young and did not know of the arguments for the wave theory, so he did not at first specify the color by its wavelength; instead he specified the color by its deflection angle in a "standard prism" made of his very pure flint glass. He tested many light sources for this work and noticed that one type, using alcohol and sulfur flames seeded with salt, produced a sharp bright yellow/orange line that was always at the exact same deflection when seen through his standard

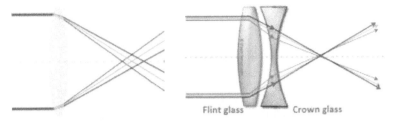

Fig. 2.   Left: Chromatic aberration (CA) of a lens. Right: CA is reduced by using a convex lens of Flint glass matched with a concave lens of Crown glass so that when their dispersive powers cancel, there will still be some net focusing because the indices do not cancel in the same proportion.

prism. (We now know this line was produced by sodium atoms from the salt; in fact, as we will see later, it is two lines very close together, now called the sodium doublet *D* lines.) If he could find more of these lines, he could use them as color markers rather than rely on a particular reference prism. Fraunhofer decided to see if sunlight had this line, too. He modified his apparatus to use a beam of sunlight through a slit attached to a window of his lab, and made one of the greatest discoveries of the 1800s, as detailed in the epigram to this chapter. He saw a very large number of "vertical lines of different thicknesses, *darker* than the rest of the spectrum." Fraunhofer counted 574 lines which he painstakingly and accurately drew. Some are shown in Fig. 3, which is his published sketch of the lines released as a German postage stamp commemorating the 200th anniversary of his birth.[1]

Fraunhofer quickly carried out some basic observations of the lines. First he showed that they are not produced by the atmosphere of the Earth through which the light from the Sun and stars must pass to get to us, because bright stars, such as Sirius, also show lines but not with the same arrangement. He concluded that the configuration of lines tells us something about the star that produces them, which began the still thriving field

---

[1] As pointed out earlier, the lines had always been there of course, just smeared out by imperfections. Wollaston in 1802 had noticed five or six vague divisions in the solar spectrum, but his resolution of dispersion was clearly not good enough. Fraunhofer's lines were not really anticipated by any previous observation.

Fig. 3.   The Fraunhofer lines in the solar spectrum, reproduced on a postage stamp.

of stellar spectroscopy. Fraunhofer converted to the wave theory in the early 1820s, in which it is natural to say each line occurs at a *definite wavelength*.

The next question was why were the lines *darker* than the surrounding region of the spectrum. First, recall that the lines were originally discovered while searching for a counterpart to the *bright* line seen in one of Fraunhofer's lamps. Fraunhofer soon discovered that when spectra of various lamp sources were examined with care, they showed bright lines, too, some of them at the same wavelengths as dark lines in the solar spectrum. It was eventually realized that the dark lines occur at wavelengths where atoms or molecules *absorb* radiation and the bright lines where atoms or molecules *emit* radiation, a point to which we will return later in the chapter.

For Fraunhofer and others after his discovery, these lines provided hundreds of precise *wavelength* markers as references; they were independent of a particular reference prism and would be the same for everybody. Important as the lines were for measuring dispersion, it was only a hint of the dramatic impact they would have on the entire development of physics, from the study of stars to the study of atoms. A set of lines is associated with a given atom or molecule, and when the lines of a given element seen on Earth are also seen in the solar spectrum, they tell us that element exists in the Sun, too. In general, these lines offer a means of telling whether a type of atom or molecule exists in a sample, and if so, how much. The study of *spectral lines* (a term we will use interchangeably with the term *Fraunhofer lines*) became important not only in physics but in chemistry, biology, and the Earth sciences.

All of this became possible when Fraunhofer found a way of measuring the positions of the lines more accurately than his prism allowed. Unknown to him, Thomas Young had briefly investigated a technique tailor-made for his purpose, what is now known as the *grating spectrometer* or the diffraction grating.[2] In 1820, Fraunhofer rediscovered and exploited this technique, and from this time onward, the grating spectrometer became the workhorse of spectroscopists and is still being used and improved. His first gratings were made by stringing fine parallel wires close together to produce "slits", which were the narrow spaces between the wires.

Figure 4 provides an explanation of a grating and how it produces dispersion. (Caution: This explanation is not for everybody, and some readers may want to skip this paragraph and simply accept that gratings enable such wavelength resolution as seen in Fig. 4(c).) As a preliminary, Fig. 4(a) shows light shining through an opening and landing on a screen, forming a single diffraction peak at the center as in Figs. 6 and 7 of Chapter 7 (though in the present case the subsidiary maxima are assumed too faint to be visible). Now comes the grating part, in Fig. 4(b). If we fill the opening with a grating having equally spaced slits, we still get the narrow diffraction peak at the center, but we also get interference peaks on each side (just the first two are shown in the figure). These occur where the wavelets from all the slits have a common phase, which happens when the paths of adjacent wavelets reaching the same place on the screen differ by an integral number of wavelengths. When there are many colors present, as in the white light source of Fig. 4(b), the side peaks are spread out because the wavelength of each color reinforces at a different place on the screen; in other words, there is dispersion. The dispersion is caused by the effect of color on interference

---

[2] After David Brewster reported in 1814 that illuminated mother-of-pearl (a shell that has a large number of parallel grooves) and other natural substances produce unusually pure colors separated by dark intervals, Young in 1817 had pointed out that several parallel grooves etched in a plate should produce a more intense interference pattern in reflected light than two such grooves, and should also be much sharper, but he had not followed up on this idea.

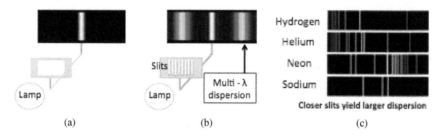

Fig. 4. (a) Light passing through a single opening creates a single-slit diffraction pattern on a screen, as was seen also in Figs. 6 and 7 of Chapter 7. Only the bright central maximum is visible and the horizontal scale is shrunk for viewing. (b) The opening has now been filled with a grating of eight equally spaced slits. White light is used, but the wavelets from all the slits produce a maximum at the center regardless of color since they have traveled approximately the same distance. However, the different colors travel different distances to the other maxima in order to maintain a whole number of wavelengths difference from each slit. This dispersion in color is much greater for 1,000 slits, as in (c), where lamps of different gases have been used and just a single side peak is displayed.

rather than on the refractive index of a prism. Figure 4(b) shows how the colors are dispersed for eight slits (there were many more in Fraunhofer's grating). We can increase this dispersion by making the slits narrower and closer together, and increase the wavelength resolution by increasing the width of the entire grating. In this way, the dispersion and resolution can be made much larger than for a prism, making it possible to see a spectrum with high resolution, as shown in Fig. 4(c) for the spectra of some elements including that of hydrogen which we will study later.

Fraunhofer's second generation of gratings was created by etching parallel grooves on glass with a diamond. With this grating, he saw that the bright sodium line that led to his discovery of the solar lines, visible in Fig. 4(c), is in fact two lines, now called a doublet, separated by only a thousandth of their wavelength, a resolution he would never have seen with his prism. Diffraction gratings were thus launched.[3]

One of the many applications of the Fraunhofer lines seen with the diffraction grating is the measurement of the Doppler effect, named after

---

[3] Both reflection and transmission gratings were used, and subsequent developments included shaping each groove to concentrate the reflection at a particular angle, called a "blazed" grating. This was coupled in modern gratings with making the grooves even closer together and the area of the grating larger.

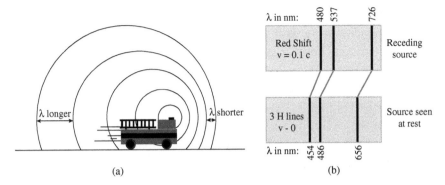

Fig. 5. (a) Illustration of the Doppler effect with sound waves emitted by a moving fire truck, producing a higher frequency detected in front and a lower frequency and longer wavelength behind it. The same effect applies to the detected frequency and wavelength of a source of light moving toward or away from an observer. (b) The wavelengths of three spectral lines of hydrogen atoms at rest, compared with the observed wavelengths from a distant galaxy receding at a speed of one-tenth the speed of light, showing the Doppler shift to longer wavelengths.

Christian Doppler (1803–1852). Doppler predicted a shift in the observed frequency of light (and hence the wavelength of a Fraunhofer line) due to the motion of the light source relative to the observer. The Doppler effect applies to any type of wave, be it sound, radio, radar, light, etc., and is illustrated in Fig. 5. The frequency shift is to higher or lower frequency depending upon whether the motion is toward or away from the observer, as shown for sound waves in Fig. 5(a). The same is true for light, and the Fraunhofer lines provide a convenient way to measure the shift. In 5(b) a source emits light of wavelength λ when seen in its rest frame, but when moving with speed V it appears to have a wavelength shift $\lambda(V/c)$ toward longer λ for V directed away from the observer. (This formula is valid only when V is much less than the speed of light c; otherwise a relativistic formula must be used.).

The Doppler shift of spectral lines of light from a star is useful in judging the velocity of the star relative to us. In fact, Doppler himself used it to show that most double stars are not two stars that accidentally lie close to the same line of sight but instead usually are two stars revolving about each other, which have different velocities along the line of sight to the Earth causing the spectral lines of one star to be shifted relative to the other, the shift changing as the stars revolve.

The Doppler effect is especially interesting when stars are in distant galaxies receding from us at speeds approaching that of light. The redshift of all the spectral lines in such galaxies is the basis for believing in an expanding universe. Figure 5(b) illustrates the redshift in the lines of hydrogen atoms contained in a distant galaxy receding from us at a speed of one-tenth the speed of light.[4] (These hydrogen lines are part of the Balmer series which was shown in Fig. 4 and will be discussed further a few paragraphs hence.) In 1929, Edwin Hubble showed that the redshift in the light from each distant galaxy — and hence its recessional speed — is proportional to its distance from us, known as Hubble's Law and illustrated in Fig. 6. The distance to each galaxy was determined by the observed brightness of Cepheid variable stars within it. (Cepheid variables had been shown to be "standard candles" by Henrietta Lovett, as explained at the end of Chap. 5, so that their apparent brightness in a galaxy measures how far the galaxy is from us.)

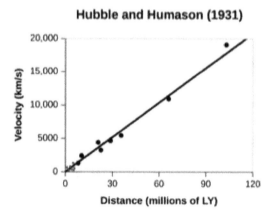

Hubble and Humason (1931)

Fig. 6. The redshifts of wavelengths, and hence the recessional velocity, get larger the further away the galaxy, indicating that the universe is expanding.

---

[4] For the really distant galaxies, general relativity tells us that it is more accurate to think of the redshift as due to the expansion of space rather than the Doppler shift; that is, the wavelength of light from those galaxies has been stretched in step with the stretching of space that has occurred during the billion or so years while the light is traversing the distance to here.

The Doppler shift illustrates one of the many important applications of the Fraunhofer lines, but an obvious question is why does an element emit (or absorb) only certain lines in the first place; for example, the atomic hydrogen lines we've just seen? Though there were many theories, the correct explanation eluded scientists until the 20th century. The pathway to the correct idea lay in cracking the code posed by the very simple spectrum of atomic hydrogen shown in Fig. 4(c) and 5(b). The person who led the way down this path was a very unlikely trailblazer.

Johann Balmer (1820–1898) was a teacher of mathematics in a girls' secondary school in 1880 and an occasional lecturer at the University of Basel when he became aware of the problem of interpreting the spectral lines of atomic hydrogen. He was 60 years old with no publications in physics to his name. Fortunately perhaps, he did not know about the prevailing theories of spectral lines, so he had no preconceptions. By some method — we do not know how — in 1885 he finally arrived at a relation giving the wavelengths of the observed hydrogen lines in terms of an integer n that labels each line:

$$\lambda_n = \lambda_o \, n^2/(n^2-2^2), \quad n = 3, 4,... \quad (1)$$

where $\lambda_o = 0.36456$ microns and n is any integer greater than 2 with λ_n the observed wavelength corresponding to that integer. Rewriting in terms of the frequency[5] $\nu = c/\lambda$:

$$\nu_n = \nu_o \, (1/2^2-1/n^2), \quad n = 3, 4,... \quad (2)$$

which predicted all the H lines then known, and some which soon were measured in stars by colleagues at Basel working in the ultraviolet.

The agreement of this simple formula with the data was astounding, and it was immediately accepted as the correct series formula, the first that had been obtained for any of the thousands of Fraunhofer lines. Then, Lyman discovered a separate series of hydrogen lines entirely in the

---

[5] We use the Greek letter ν to denote frequency from this point on, rather than the letter f as in earlier chapters.

ultraviolet and Paschen found another series in the infrared. Johannis Rydberg showed all these lines could be included in a single formula as follows:

$$\nu_{nm} = \nu_o \, (1/m^2 - 1/n^2) \quad \text{with } \nu_o = cR_H \tag{3}$$

where $R_H = 1.0968 \times 10^{-7}$ per meter is the *Rydberg constant* already known quite accurately from experimental measurements of the Balmer lines by the 1890s. Also, $m = 1$ for the Lyman series, $m = 2$ for the Balmer series, $m = 3$ for the Paschen series, and higher integer values of m presumably indicated series too far in the infrared to be observed at that time. This simple and exact relationship begs for an equally simple explanation, but it would not be discovered until 20 years later when Niels Bohr developed the first quantum theory of the hydrogen atom. His theory would predict not only the form of the above equation but also the actual value of R_H. All of this we will present in Chapter 10 when we are better prepared.

We will now shift gears and point out that another noteworthy property of the Fraunhofer lines is they naturally occur at visible wavelengths in the atmospheres of the Sun and stars where they were first seen, but *not* in the atmosphere of the Earth. We can generate the lines on Earth from atoms of most elements by flames, electric discharges, etc., but these elements do not occur as separate atoms in our atmosphere; instead they are bound in solid chemical compounds in the Earth or in molecules such as oxygen, nitrogen, and water vapor which are in our atmosphere but do not absorb much in the visible. This transparency of the atmosphere is of course very important for us because it enables us to see things even miles away, and is likely the cause of us and other animals evolving eyes that are sensitive in this wavelength region.

It is a different story at infrared (IR) and ultraviolet (UV) wavelengths, the invisible regions of the spectrum first discovered in 1801–1803 as we mentioned in Chapter 6 and alluded to elsewhere. In those spectral regions Fraunhofer lines indeed exist in our atmosphere, and for some wavelengths the absorption lines are so dense that there is no transmission at all even over very short distances. The absorption spectrum in the IR, which will be of interest to us in what follows, is shown in Fig. 7

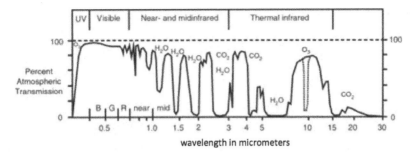

Fig. 7. The absorption in the atmosphere due to the presence of Fraunhofer lines of molecules. Note the small absorption in the visible, but large absorption in the IR, where in some places it is continuous because of the high density of Fraunhofer lines. Roughly speaking, the reason that molecules absorb in the IR is that they do so by the internal motion of their constituent atoms which have resonant frequencies that match the frequency of light waves in the IR but not in the visible. Dense Fraunhofer lines also appear in the UV where they are produced by the motion of the much lighter electrons inside the molecules which have high frequencies that match those of light waves in the UV.

together with the absorption in the visible and UV for comparison. In the IR, water vapor is the largest absorber, but carbon dioxide CO_2 and other *greenhouse gases* play a large role as well.[6]

Atmosphere absorption in the IR has received much attention in recent decades in the context of *global warming*. We will sketch the basics here. What determines the average temperature of Earth? It is the balance between the heat absorbed by Earth from sunlight and that lost by radiation from Earth into space. The Earth is warmed by radiation from the Sun, which is mainly visible and thus can penetrate the atmosphere. Some sunlight is reflected back into space by our atmosphere and by the Earth, and the remainder (about 70%) is absorbed by the Earth and heats it up. The heated Earth radiates back into space, mainly in the IR, and reaches an average temperature at which the heat lost to space just

---

[6]They are so named because they absorb heat radiated by the Earth and prevent it from escaping to space, warming up the Earth and acting like a greenhouse roof, which causes the inside to heat up. However, the roof warms the interior mainly by reducing the cooling from convection, though it does warm a bit by absorbing the outgoing IR radiation as well. So, the name is a bit of a misnomer.

balances the heat gained from the incoming solar radiation. (Other sources of heat such as radioactivity, volcanism, tidal friction, etc. are negligible by comparison.) If there were no atmosphere, this average temperature of the Earth's surface can be calculated and would be about −18°C (−1°F).

There is an atmosphere, however, and because of those Fraunhofer absorption lines in the IR where most of the radiation back into space would occur, some of that outward flowing radiation is absorbed and reradiated back down to Earth rather than escaping to space, causing the Earth's surface to be hotter. So the Earth heats up to an average surface temperature of 15°C (59°F) to compensate and produce enough radiation into space to equal the incoming radiation. Greenhouse gases such as CO_2 and methane have been increasing in our atmosphere due to fossil fuel burning and other causes, creating more absorption of outgoing radiation by our atmosphere and hence an increase in global average temperature. Figure 8 shows the measured CO_2 concentration in the atmosphere, which has clearly increased in the past 70 years. Figure 8 also shows the global mean temperature over the past 140 years, which has exhibited up and down variations over much of the time, but an average rise for the past 40 years. There are competing factors also affecting the radiation balance such as sea ice reflection, variation in water vapor concentration, etc. The well-known consensus among atmospheric scientists as of this writing is the Earth is warming up due mainly to increasing CO_2 in the atmosphere.

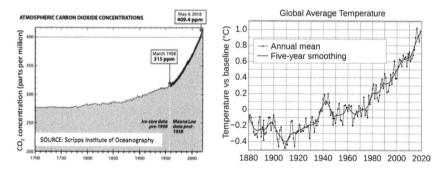

Fig. 8. Plots of CO_2 concentration in the atmosphere and of global average temperature in recent decades.

Let us turn now to still another important feature of the Fraunhofer lines that was not appreciated when they were first seen. The wavelength of a line is *fixed in value*, having the same value on Earth, on a star, or presumably anywhere in the Universe. Suppose we wish to have a standard measure of length so that, for example, people building an aircraft engine in Europe will make it to the specified length to fit on the wing being assembled in the U.S. In other words, we want to agree on the same standard of length, such as the meter stick that was held in Paris for decades. People would have to go to Paris and try to compare their version to the standard one, and this would clearly be difficult. Also, how accurately could the stick length be maintained against variations in temperature, etc? A much more accurate and convenient way is to measure the stick by counting the number of wavelengths of a convenient Fraunhofer line that fit within the length of the stick. In fact, one can just define the length of a meter as containing a specified number of wavelengths of the line, and then those of us not living in Paris wouldn't have to go there, pleasant as it usually is, to have our very own standard. We just have to agree on which Fraunhofer line, and then everyone can do consistent measurements, wherever they are, even residents of a distant planetary system if we are in communication with them. Certainly, the aircraft engine in Europe will fit in the wing made in the U.S. when the crucial size measurement in each case is referenced to a Fraunhofer line.

The instrument for doing this is called an *interferometer*. The name refers to measurements that utilize wave interference and employ Fraunhofer lines when needed for exceptional accuracy. A sketch of one version, called the Michelson interferometer, is shown in Fig. 9. An incoming beam of monochromatic light is split into two beams by M, a partially silvered mirror that transmits about half of the beam and reflects about half; one beam traverses the reference arm, reflects from a reference mirror M1, and returns, the other beam traverses the measurement arm i.e. the meter stick or whatever we want to measure, and returns by reflection from a moveable mirror M2 to greet the reference beam at the original beam splitter M where the two beams recombine and are detected. If the round trip distance of the M1 beam differs from that of the M2 beam by an integral number of whole wavelengths, we will see a bright fringe, if M2 then moves so its round trip distance changes by a half wavelength, we will see a dark fringe. We can measure the length L of the stick (or other desired object) in terms of wavelengths by counting the number of bright fringes that are created as we move the measurement arm the length of the stick.[7] This is exactly what Michelson did in Paris in 1892–1893 using the narrow red line of Cadmium as the source of light for the measurement. This line served as the standard of length, revised many times since by ever more precise measurements with other Fraunhofer lines. Of course, for many objects one doesn't need such great accuracy, but it is amazing how many measurements do today. Nowadays there is a similar unchanging standard for many other quantities, such as atomic clocks for time intervals, etc.

A Michelson interferometer had already been used in 1887 for one of the most famous experiments of all time, the Michelson–Morley experiment, which showed that the speed of light measured by an observer is independent of the observer's velocity relative to the then assumed aether — a total surprise to physicists at the time. We will say very little about this experiment except to note that no change in interference pattern was observed as the interferometer of Fig. 9 rotated, causing the perpendicular paths relative to the supposed aether to change direction; nor was any change seen throughout the year as the Earth presumably changed its direction of motion relative to the aether. This result helped to kill the aether hypothesis and establish the speed of light c as a universal constant,[8] a fact we already alluded to in Chapter 4 when we described the first measurement of c.

Our final example of an interferometer application is quite up to date: the detection of gravitational waves first reported in 2015 by LIGO (Laser Interferometric Gravitational wave Observatory). These waves were

---

[7] Actually, a spectral line from a discharge lamp such as what Michelson used does not have a *coherence length* long enough to change the length of one arm by more than a few mm, let alone a meter, before well-defined interference is lost. So the measurement is more complicated! Coherence length is discussed in connection with lasers in Chapter 10.

[8] The constancy of the measured value of c independent of the source or the observer is a postulate of the Special Theory of Relativity developed by Albert Einstein in 1905. In Einstein's paper that year, he used light beams extensively to illustrate measurements of simultaneity, etc. Unfortunately, a discussion of Relativity is beyond the scope of this book.

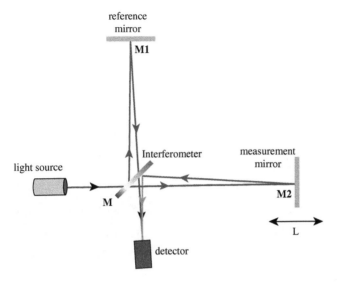

Fig. 9.   A Michelson interferometer.

predicted by Einstein's general theory of relativity in 1916. They are produced by the acceleration of any massive object, but it takes compact
objects such as black holes or neutron stars orbiting about each other and
spiraling inward to produce observable signals. The first detected signal had
as its distant source the last stage of two inspiraling black holes swallowing
each other up over a billion years ago. By the time the waves had reached
Earth, they were feeble and it took heroic efforts to detect them. Since then
other gravitational wave signals have been seen, such as two orbiting neutron stars pulling each other together, which also released a light signal at
the same time which could be correlated with the gravity wave signal.

The first gravitational waves were detected by two Michelson interferometers (the I in LIGO), each similar to Fig. 9, but built on a large scale
with 4 km arms. One is based at Hanford, Washington, and is shown in
Fig. 10(a); the other is in Livingston, Louisiana. At either site the light
source for the interferometer is a laser that operates on a Fraunhofer line
of a doped crystal. (Patience. We will describe lasers in detail in Chapter
10.) The laser produces a narrow beam (but wide enough to be consistent
with the diffraction limits we studied in Chapter 7) that is split into two
beams that travel the 4 km length of each arm and return, where they

(a)

(b)

Fig. 10.   (a) The LIGO interferometer at Hanford, WA. (b) The first gravitational wave detection: simultaneous signals at Hanford (H1) and Livingston, LA (L1). Vertical axis measures fractional change in length of arm = 5×10^{-22} per division. Horizontal axis measures time, 0.05 sec per division. *LIGO website.*

recombine and interfere as described in Fig. 9 above. However, the length of each arm is effectively multiplied to 1,200 km by placing highly reflecting mirrors at each end of both arms so that each laser beam bounces back and forth about 300 times before recombining. Only signals at Hanford and Livingston that are correlated with each another are retained; a picture of the first discovery showing the correlated signals is displayed in Fig. 10(b). The wiggles in each signal reveal the motion of the back holes as they orbit about each other, picking up speed and hence shortening the wiggle period as they spiral inward to merge. A later interferometer outside Pisa, Italy, helped in the detection of the neutron star merger, and has been participating since.

So, what are these waves and how are they detected? In general relativity, gravity is the result of the warping of space around a massive object such as the Sun, which causes objects nearby, such as the Earth, to move in an orbit about them. A gravitational wave is a warping of space which travels as a wave somewhat similar to the warp of the surface of a pond which travels as a ripple. An exact description is clearly above our level here; however, briefly the net effect of such a wave passing the interferometer in Fig. 10 is to shorten one of the arms while lengthening the other, and then the reverse. The frequency of the reversal is the frequency of the wave. The longer the arms, the greater the effect and the more sensitive the measure. The size of the effect was a fractional change in length of one of the arms amounting to a part in 10^{20} for the first detection of gravity waves. Clearly, this required unprecedented sensitivity! It would be fun to describe the Hanford and Livingston interferometers further here, but we will simply refer the interested reader to the LIGO web page for an excellent, simplified discussion.

We began this chapter with the discovery of spectral lines by Fraunhofer, and then continued by giving some highlights of the amazing developments to which this discovery led. Fraunhofer himself recognized the importance to pure science of what he was doing, and he made many observations to get applications launched as we pointed out, but then he returned to his real love, making better glass. And for that he was honored: receiving an honorary doctorate from the University of Erlangen, made a Knight of the Order of Merit of the Bavarian Crown by King Maximilian I, which classified him as a noble, and made an honorary citizen of Munich. In 1826, he died of heavy metal poisoning, like many glassmakers of the time. He was only 39, the same age as Fresnel when he died in 1827. Fraunhofer carried most of his glassmaking knowledge to the grave, but fortunately not the knowledge of the Fraunhofer lines themselves.

Chapter 9

Light is an Electromagnetic Wave

"The agreement of the results seems to show that light and magnetism are affections of the same substance, and that light is an electromagnetic disturbance propagated through the field according to electromagnetic laws."

James Clerk Maxwell, *A Dynamical Theory of the Electromagnetic Field* (1864).

In 1820, Danish scientist Hans Christian Oersted noticed something strange while demonstrating some properties of electric currents to his lecture audience of advanced physics students. The needle of his compass, placed nearby for another part of the lecture, deflected when he turned on the electric current, then went back to its former position when he turned the current off. He repeated the procedure a couple of times and then finished his lecture. In the weeks afterward, he made a systematic study of this strange observation using larger electric currents and published the results. He knew it was an important discovery — the first observation of a link between magnetism and electricity. It would start a revolution. The study of this new phenomenon, electromagnetism, would lead to such applications as electric generators and motors and the production of radio waves. It would also reveal the amazing fact that light itself is an electromagnetic (EM) wave, and that the speed of light could be calculated from measurements with coils and charged plates. This is the central result we will build up to in this chapter.

Within a week after witnessing a demonstration of Oersted's results by Francois Arago at the French Academy in Paris, Jean-Marie Ampere devised several experiments to show that electric currents exert magnetic forces on each other. Then in the 1830s, Michael Faraday in England discovered that a changing magnetic field induces an electric current in a conducting wire coil. These results of Ampere and Faraday were all that was needed for understanding how telegraph signals could travel at the speed of light. Later, in 1860 James Clerk Maxwell devised the full set of laws of the Electromagnetic (EM) field. These laws predicted that EM waves can propagate in free space, and that light is such an EM wave. Finally, in 1880 Heinrich Hertz demonstrated the generation of these waves at radio frequencies that quantitatively agreed with Maxwell's EM equations. The way was clear for applying these equations to light as well. This is a huge field and we can only touch upon the main ideas on the road to understanding light as an EM wave. Also, it is a difficult subject for many of the readers of this book; for example, all of Maxwell's equations require a knowledge of calculus. Our aim, then, will be to express the main ideas in an understandable form without calculus, and tell the interesting narrative leading to the remarkable conclusion that light is an EM wave. Along the way we will become familiar with electric forces so that we can understand how they act inside the atom in Chapter 10. Finally, as promised way back in Chapter 2, we will at last explore the physical basis of refraction and get a glimpse at why visible light moves slower in transparent materials. Also, we will explain dispersion and see why white light breaks up into colors when refracted. We will thus close the loop on the first 3 chapters of the book. It all has to do with light being an EM wave!

What were the precursors to Oersted's discovery in 1820? Two separate fields of study had grown up, magnetism and electricity, which go back to ancient times. Magnetic materials occurred naturally in the form of lodestone, which in turn could induce magnetism in iron and other materials. All magnets have a north and a south pole; if free to rotate, the magnet aligns with its north and south poles in the approximate direction of the geographic north and south of the Earth, which is the basis for the magnetic compass. Like poles repel, unlike poles attract. Electric materials also occurred naturally, amber for example, but they had to be rubbed to charge them and make them pick up pieces of paper, etc. There were positive and negative charges, and similarly to magnetic poles, like charges repel and unlike charges attract. They do not come in opposite pairs on the same body the way magnetic poles do. For example, a glass rod acquires a net positive charge when rubbed with a silk cloth while the silk acquires a net negative charge in the process.

In the years just before Oersted, more progress was made on electricity than on magnetism. In the 1790s, Charles Coulomb showed that the force on a charge q due to a second charge q' varies directly as the product of the magnitudes of the charges, qq' and inversely as the square of the distance r separating them. The force is repulsive (directed away from q') if the charges have the same sign, and is attractive if they have opposite signs.

$$F = qq'/4\pi\varepsilon_0 r^2 \qquad (1)$$

where the value of the constant ε_0 depends upon the units used. We will come back to it later. When one or more charges q' are present, it is customary to define an *electric field E* at any point in space around them, such that if a test charge q is placed at that point, the total force of all the charges on q is $F(total) = qE$ so:

$$E = F(total)/q \qquad (2)$$

In Fig. 1 we show the electric field in the vicinity of two equal charges of opposite sign, where we have taken account of the important fact that

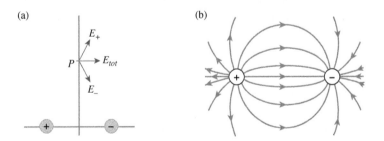

Fig. 1. (a) A calculation of the electric field E at a point due to two equal charges of opposite sign, taking account of the vector nature of E. The vertical components of $E+$ and $E-$ cancel while the horizontal components add. (b) Calculation done like the previous one but for all points in the entire region about the charges. The field directions form lines of E, which are more closely spaced where the field is larger near the charges.

F and *E* are vector quantities having magnitude *and* direction. The figure shows lines of *E*, which by convention get closer together the larger the magnitude of *E*. This notion of an electric field has proven to be a very useful concept, and will play a central role later in our discussion.

In 1800, Alessandro Volta, while exploring Galvani's observation of the effect of electric charges on animals, made a device called a Voltaic *pile* pictured in Fig. 2 that could produce a continuous flow of electricity. This was essentially the first battery, consisting of a stack (hence the name pile) of silver and zinc disk pairs; the silver and zinc disks of each pair were separated from each other by a paper or cloth soaked in brine. Early improvements included replacing silver with copper, and brine with dilute sulphuric acid. The copper terminal was determined to be the positive electrode and the zinc terminal the negative one by their interactions with known charges, eventually becoming, by convention, the accepted definition of positive and negative.

The pile produces what came to be called an electromotive force, emf, or voltage between the terminals, and a wire connecting the terminals carries a continuous electric current, called *I*, defined as the *quantity of charge per second* moving past a cross-section of the wire, the same as the quantity transferred in a second from one terminal to the other. The voltage drop *V* along the length of the wire is defined as the electric field *E* in the wire times the length *L* of the wire, or $V = EL$.[1] This was the state of affairs when Oersted made his great discovery in 1820.

Hans Christian Oersted was born in the small Danish island town of Rudkoebing to the village apothecary Soeren Christian Oersted and his wife Karen Hermandsen. There was no school in the town, but he and his brother Anders managed to get an informal education from various townsfolk: a wigmaker and his wife who taught them German and how to read and write Danish, the town surveyor who taught them mathematics and drawing, and the mayor who taught them English and French. Hans Christian began to work in his father's apothecary shop at age 12, where he became interested

---

[1] For completeness we mention that Ohm's law $V = IR$, where R is the resistance of the wire, and other rules of electric circuits were developed. Soon electrolysis was discovered, for example, water producing hydrogen gas at the negative terminal and oxygen at the positive one. The huge field of electrochemistry was born and these and related studies occupied scientists throughout Europe after 1800.

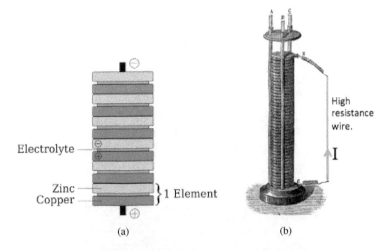

Fig. 2. Voltaic pile, the first type of battery. (a) Detail of arrangement. (b) An early example of use in a circuit.

in medicine and science. He and Anders then attended the University of Copenhagen, he majoring in the sciences of medicine, physics, and astronomy, and Anders studying law.[2] Hans earned his doctorate and eventually became Professor of Physics at Copenhagen.[3] After Volta's discovery he built his own voltaic pile and conducted experiments with it. Whereas the French scientists believed that magnetism and electricity were separate forces and differed from each other just as much as each differed from the force of gravity, Oersted sided with the German school which held that magnetic and electric forces had enough similarities that they must be linked in some as yet undiscovered way. He found a link in 1820.

After his initial observation of that link during the lecture mentioned at the beginning of this chapter, Oersted developed a larger, more

---

[2] Anders rose to become the foremost jurist in Denmark and a Minister of State.

[3] At almost the same time, Hans Christian Andersen, a poor boy of 14, arrived in Copenhagen and somehow went to Professor Oersted's house and introduced himself, impressing everyone in the house with his charm, so that they saw to his upkeep. A long friendship ensued. Andersen would refer to Oersted as the Great Hans Christian and to himself as the Little Hans Christian. Oersted's youngest daughter Matilda, became Andersen's protégé and on his death she was bequeathed the manuscripts of his world-famous stories.

Fig. 3.   Hans Christian Oersted.

powerful pile and other improvements and undertook a very careful study illustrated in Fig. 4. A wire that could be connected to a voltaic pile by a switch was oriented north–south and a compass needle that naturally pointed in a direction parallel to the wire due to the Earth's magnetism was placed beneath the wire. When the switch was closed so that current flowed in the wire, the compass needle deflected toward the east–west axis perpendicular to the wire as seen in Fig. 4(a). The deflection was in the opposite direction when the wire was arranged to be beneath the needle. Oersted arranged an apparatus so he could test all relative orientations of wire and needle, and obtained results summarized in Fig. 4(b). In essence, an electric current in a wire creates a circular *magnetic field* around the wire as axis; the direction of the field reverses if the current is reversed. (For the present, we take the following as a working definition of a magnetic field at a point in space: it has the direction that a compass needle points when placed there and a magnitude proportional to the twist or torque it exerts on the needle to align it.) Placing various non-magnetic materials between the wire and the needle, such as wood, glass, or water, had little effect. These are simple if profound observations and although many scientists had seen hints of the effects in the 20 years since the voltaic pile had been invented, they had not followed up on them systematically until Oersted.

Oersted described his observations in a famous four-page paper, written in Latin, circulated to friends and science groups in Europe and then translated into several languages and published in journals of those countries. There were literally hundreds of notices by scientists following up on these discoveries in the next two years. Oersted himself received a profusion of congratulations, honors, and awards; he was made a fellow of many learned societies, awarded the Copley Medal by the Royal Society, and given a prize of 3,000 gold francs by the Institut de France. Oersted did not stop his wide-ranging scientific activities, and in 1825 he made the first ever isolation of the element aluminum.

Among the scientists inspired by Oersted's discovery, Andre-Marie Ampere — Professor of Mathematics at the Ecole Polytechnique in Paris — was the one who eventually emerged with the correct path forward. Francois Arago, already known to us from Chapter 7 as a champion of Fresnel, had witnessed a demonstration of Oersted's discovery in Geneva and gave a demonstration himself at the Ecole. Seven days later, Ampere presented a paper that sketched most of the ideas that he would later develop into careful experiments showing that electric currents alone could produce a magnetic effect on each other, and in fact could be the basis of the magnetism in actual magnets. However, Jean-Baptiste Biot, another researcher at the same institute, in company with the majority of his colleagues, held a contrary view, that the current in the wire created temporary little magnets while it flowed, which were the source of the needle deflection. So the stage was set for the rival groups at the same institute to carry out experiments they imagined would support their own views. Biot would do some useful experiments, and even have one law of interactions named after him, but in the end Ampere's view was the one that would prevail.

Andre-Marie Ampere was born in 1775 in the small French village of Poleymieux near Lyons. He was educated by his family in most subjects and a private tutor in mathematics for which he showed exceptional aptitude. His interests ranged widely, and included ancient languages, botany,

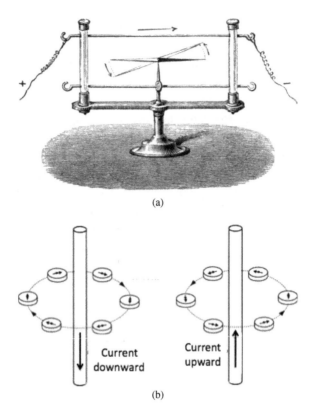

(a)

(b)

Fig. 4. Oersted's discovery: (a) Compass needle points N–S along Earth's magnetic field; wire directly above it is arranged N–S too. When electric current flows in the wire, the compass is deflected toward E–W, reversing deflection when the current is reversed or when the current flows through the wire beneath the compass needle. (b) Compass needles arranged in a circle about current carrying wire. When current flows downward, the needles point as shown. When current is reversed, the needles reverse their direction, too. Earth's field has been canceled out with a magnet not shown.

astronomy, and physics, as well as the construction of instruments. At age 22 he moved to Lyons to earn a living on his own, which he did by private lessons in mathematics, physics, and chemistry and even setting up a small instructional laboratory in his home. In 1804, he was offered a position in mathematics at the Ecole Polytechnique in Paris and then made a member of the Academie Francais in 1814 for his work on the theory of games. There he was soon to become an early and ardent supporter of

Fresnel's wave theory of light. Then came Arago's demonstration of Oersted's discovery to the Academy.

Ampere's first experiment, shown in Fig. 5(b), verified his idea that a force exists between two current-carrying wires. He found that parallel wires attract each other if their currents are in the same direction and repel each other if their currents are in opposite directions. For wires each of length L separated by a distance r, carrying currents I_1 and I_2, the correct expression for the force is

$$F = \mu_0 I_1 I_2 L/2\pi r \tag{3}$$

where μ_0 is a constant whose value depends upon the units chosen for the other quantities in the equation. If we compare this equation with Eq. (1) for the Coulomb force between two charges, it can be shown that $1/\mu_0\varepsilon_0$ has units of velocity-squared. Can you guess what velocity it turns out to be? It will be revealed here shortly.

Ampere then followed up on this result with other experiments to find how the force between two wires varies with the relative orientation of the wires, often connecting a wire to the rest of the circuit by a pool of mercury (which is electrically conductive) at each end of the wire to allow freedom of movement while maintaining electrical contact.[4] Ampere's plan was ingenious and his results were described years later by James Clerk Maxwell (whom we will meet in due course) as "perfect in form, and unassailable in accuracy."

These results can be expressed in many ways. Ampere's way was to find directly the force between two current-carrying wires as we have seen. Our way henceforth will be to utilize the *magnetic field B* (often called the magnetic induction) which we introduced with Oersted's discovery. It is

---

[4] Although Ampere had worked with apparatus since childhood, he was always more adept at designing and interpreting experiments than in constructing and performing them, unlike many of his famous predecessors. He would give the plan of an experiment to a member of the laboratory at the Academy who would then build and often carry out the experiment under Ampere's guidance. There is an amusing anecdote in Oersted's diary about when he visited the Academy at the height of the rivalry with Biot. Ampere endeavored to show Oersted how his apparatus worked and in the process misaligned several parts so that he had to call a technician to set it right and do the demonstration.

(a)   (b)

Fig. 5. (a) Andre-Marie Ampere. (b) Ampere's experiment showing attraction or repulsion between two current-carrying wire loops. If the currents in the two wires are in the same direction, the wires attract each other; if the currents are in opposite directions (as is the case in the figure), the wires repel.

the magnetic analog of the electric field E. B has a direction and a magnitude at each location in space; the direction is the same as a compass needle would point if placed at that location, and the magnitude tells us how strong the field is there. We will illustrate with a simple example the field surrounding a long straight current-carrying wire as shown in Fig. 6. B points along a circular path about the current as axis as we showed earlier with the compass needles in Fig. 3(b).[5] The magnitude for B in this case is given by

$$2\pi r B = \mu_0 I \tag{4}$$

where I is the current, r is the distance from the wire, B is measured in tesla, and μ_0 is the constant shown already in the equation for the magnetic force.[6]

---

[5] By convention, if we align the thumb of our right hand with the electric current and curl our fingers, then the fingers point along B. We don't need a compass needle.

[6] Although we won't need it for our work, it is interesting to know how the force exerted by B on another current carrying wire is determined. We can get an idea from Ampere's result for the force between two current carrying wires shown in Eq. 3 and depicted in Fig. 5(b), which gives us the force (magnitude and direction) exerted on one current I_1 by the field B of another current I_2 when I_1 is perpendicular to B, and B is determined by Eq. (4) and depicted in Fig. 6.

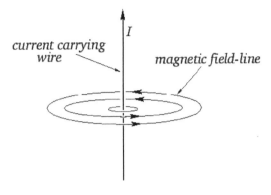

Fig. 6   The magnetic field around a current-carrying wire by Ampere's Law.

Equation (4) is a special case of what is called Ampere's Law. In the general case of an arbitrary closed path, not necessarily a circle, with a total current I passing through it, but not necessarily having the symmetry of Fig. 6, the field B will usually vary in magnitude and direction from point to point along the path but amazingly satisfies a similar equation to Eq. (4). This unfortunately requires calculus to express. We will represent it by

$$Loop\ B = \mu_0 I \qquad\qquad (5)$$

where *Loop B* is used in place of the calculus expression. Stated in words, *Loop B* is the average value of the component of B along the loop path multiplied by the distance around the loop and reduces to the left-hand side of Eq. (4) for the situation of Fig. 6. We will refer to Eq. (5) as *Ampere's Law*, but needless to say we will not use it for any calculations.

At the same time that Ampere performed his earliest experiments with a straight wire as in Fig. 5(b), he used a compass needle to map the magnetic field of a "solenoid", a current-carrying wire in the form of a helix pictured in Fig. 7(a). He mapped the field with iron filings sprinkled on paper as shown in Fig. 7(b), and saw that the field resembles that of a bar magnet shown similarly by iron filings in Fig. 7(c). He then proposed a remarkably prescient idea, that an iron magnet actually consists of a solenoid current flowing around it. This idea was generally accepted after Fresnel, then a colleague of Ampere's at the institute, suggested that a similar field would be produced by the sum of permanent circular currents flowing in microscopic domains, as illustrated in Fig. 6(d).

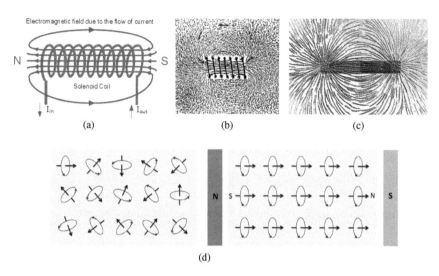

Fig. 7.   (a) The field B of a current-carrying solenoid. (b) The field of a solenoid revealed by iron filings on a flat piece of paper at the midplane. (c) The field of a bar magnet revealed by iron filings as in (b), looking very similar to the filings in (b). (d) A magnet actually consists of magnetized domains, pointing in random directions until lined up by an external magnet.

This view is similar to the one held today, that microscopic magnetic domains are the little magnets in magnetic materials, lining up in an external field to produce an overall magnetic effect. In "soft" materials such as iron this magnetism is lost when the external field is removed and the little magnets revert to pointing in random directions, whereas in "hard" materials in which the domains resist changing their orientation, such as steel, a permanent magnet is formed.[7]

The idea of a field having a real existence was reinforced by observations such as in Figs. 7(b) and (c), in which one can almost "see" the field B revealed by the iron filings. This view gained more adherents as a result of what Michael Faraday discovered a few years after Ampere's work.

Michael Faraday (1791–1867) was born near London to a former blacksmith's apprentice who could afford only the most basic education. At age 14 Michael became an apprentice to a bookbinder and bookseller

---

[7]Though this view of magnetic materials has been held since Ampere's time, it was not until the 20th century that physicists finally understood through quantum mechanics *why* the atomic spins in a domain line up in certain materials such as iron.

for 7 years during which he read books such as *The Improvement of the Mind* by Isaac Watts which inspired him, and tracts on science, especially chemistry. At 20 he attended the Royal Society lectures of Sir Humphrey Davy on chemistry during which he took detailed notes, bound then into a 300-page book and then presented them to Davy. Davy was impressed, and offered him a position as Chemical Assistant at the Royal Institution in March 1813, where he remained for the rest of his life, rising to the position of laboratory director after Davy's death. In 1821, he married Sarah Barnard, a fellow member of the Sandamanian Church (an offshoot of the Church of Scotland), which played an important role in his life; he served as a deacon and as an elder for many years.

In 1831, Faraday began his great series of experiments in which he discovered what is now called electromagnetic induction. He was convinced that an electric current in one wire could induce a current in a separate wire, just as a charge induces the opposite charge in the closest part of a nearby conductor, so he set out to demonstrate it. He was partly right, but not in the way he expected, as we will see. Since no one had seen such induction before, he thought it must be feeble, so he wound two coils of

Fig. 8. Michael Faraday.

many turns closely overlapping with each other to see if a current in one of them would induce a current in the other. One of them, called the primary coil, was connected to a battery by a switch; the other, called the secondary coil, was connected to a meter to detect any induced current. When he turned a steady current on in the primary coil, the most he could generate, he saw no steady current in the secondary. However, he noticed when he flipped the switch turning the primary current on or off, he saw a sudden jump in the secondary current, which then returned to zero. Also the jump was in the opposite direction when the primary was turned off than when it was turned on. (At first he thought he must be doing a trivial thing like shaking something when putting the switch in and out, but when he disconnected the battery, the effect of switching went away.) Faraday somehow guessed that it was a *changing magnetic field* inside the secondary coil that produced the effect. So he put two coils on an iron ring to enhance any mutual magnetic effect, even when the coils were not so close to each other, as shown in Fig. 9. He saw the same induced effect, and he was now sure of its origin. He had discovered another source of an emf (i.e. a voltage) in a circuit besides a battery — a changing magnetic field.[8]

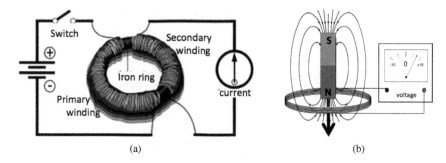

Fig. 9. (a) A circuit featuring the original coils used to study the Faraday effect. Switching the current on or off in the primary induced a current in the secondary winding, which Faraday saw was caused by changing the magnetic flux in the secondary. (b) Another way of generating the Faraday effect. A permanent magnet is moved through a coil creating a change of flux in the coil.

---

[8] Note this result was much simpler to express in terms of fields, B and E, than in terms of forces as in action at a distance theories. Nevertheless, the latter theories still had strong adherents even after Maxwell 30 years later.

By using the iron ring, which insured that the magnetic field of the primary mainly went inside the secondary, too, Faraday found that this induced emf around a closed loop is equal to the rate of change with time of the magnetic *flux* through the loop, where the flux is $\Phi_B = BA$, the strength of the field times the area enclosed by the loop. The rate of change of this quantity with time is $\Delta\Phi_B/\Delta t$, where $\Delta\Phi_B$ is the change of flux in a time interval Δt. *Faraday's Law*[9] can then be written as

$$\text{emf} = Loop\ E = -\Delta\Phi_B/\Delta t \tag{6}$$

where *Loop E*, in analogy with *Loop B* in Eq. (5), is used in place of a calculus expression and is the average value of the component of *E* along the loop path multiplied by the distance around the loop. The minus sign indicates that the *direction* of the induced emf is always such as to oppose the change in *B* that produces it; this is called *Lenz's law*. For example, in Fig. 9(a), the induced emf produces a current and associated magnetic field that is in the direction of the inducing field if that field is being turned off, and in the opposite direction if that field is being turned on.

In these details, the reader should not overlook a crucial aspect of Faraday's Law that was not appreciated by many researchers at the time: it is the first law we have encountered that can be expressed entirely in terms of fields rather than a field produced by charges (Fig 1) or currents (Eqs. (4) and (5)). Instead, in Eq. (6), a time-changing B field produces an E field. As we will see soon, in an electromagnetic wave such effects can take place far from any charge or current where only fields are present.

The Faraday effect is the basis for the transformer. As in Fig. 9(a), a primary coil creates a changing magnetic flux in a magnetic material, which induces an emf in each turn of a secondary coil, so that the emf or total voltage induced in the secondary is increased by the number of turns in the secondary coil. A step-up transformer has more turns on the secondary than on the primary, a step-down has fewer turns.[10]

---

[9] Joseph Henry in the US was close on the heels of Faraday, but his work was not appreciated across the Atlantic until later.

[10] An induction coil for producing a very high voltage is similar; a dc voltage is connected to the primary by a switch which when turned on induces a much higher voltage in the secondary because of the sudden onset of the current and associated change in magnetic

The Faraday effect can also be demonstrated by plunging a permanent magnet inside a coil as shown in Fig. 9(b), producing a changing magnetic flux inside the coil which induces an emf in the coil. The same effect occurs if the magnet is held stationary and the coil is thrust upward to enclose the magnet; it's the relative motion that counts. In the latter case, the coil is said to be "cutting the lines of B". Rotating a coil in a magnetic field is the basis for the electric generator, which converts the rotating mechanical energy of steam, gasoline, or hydroelectric turbines into electrical energy that can then be transmitted to distant places by wires. All of this development was in place by the 1840s.

People had wondered for a long time how fast an electric signal travels down a wire. From the beginning they knew it to be fast. Even though the telegraph was developed and fielded by the 1840s, the speed at which signals were transmitted was still not known. Laboratory researchers were continuing to make progress on measuring electric and magnetic quantities such as ε_0 and μ_0, which we introduced in Coulomb's and Ampere's Laws. Especially interesting was the value of $1/\mu_0\varepsilon_0$ which has units of velocity-squared as we said earlier, so in 1854, Weber and Kolrausch made a very careful measurement of this quantity by comparing the electric and magnetic forces of Eqs. (1) and (3). After measuring the charge they had placed on a conducting sphere, they discharged the sphere through a wire and measured the magnetic force produced by the discharging current. Weber and Kolrausch showed the velocity given by the square root of $1/\mu_0\varepsilon_0$ is equal to c, the speed of light, to within a few percent! Very suggestive. Weber and independently Kirchoff then showed in 1857, using Coulomb's, Ampere's, and Faraday's Laws, that a wave signal would travel down a telegraph wire with this same speed, in other words, the velocity of light. From this point on, everyone knew that there should be a connection between electromagnetism and light.

It was left to James Clerk Maxwell to complete the equations of electromagnetism and discover the connection. The first equation is Coulomb's Law, Eq. (1). The second equation expresses the fact that a magnetic field

---

flux. We will return to the induction coil when we discuss the generation of EM waves at the end of the chapter.

always consists of closed lines, as in Figs. 6 and 7(a). The third is Faraday's Law, Eq. (6). Maxwell found that the fourth, Ampere's Law, needed another term to include what is often called the *displacement current* as we will see in what follows. The amended equation became Maxwell's fourth equation. With this term added, Maxwell's four equations predict EM waves in free space as we will see.

Who was this Maxwell? He is not so famous to the general public, but to physicists Maxwell is ranked with the likes of Einstein and Newton. Indeed, when Albert Einstein was asked if one reason for his great success was that he had stood on the shoulders of Isaac Newton, Einstein corrected the statement by saying "No, I have stood on the shoulders of James Clerk Maxwell." Maxwell was born near Edinburgh, Scotland, in 1831 to Frances Cay and John Clerk, a lawyer who acquired the name Maxwell during his lifetime as the result of his grandfather's will, which ended two centuries of blood feuds, including executions and murders, between the Maxwells and the Johnstones. Although the Maxwell family was one of the most wealthy and distinguished in Edinburgh, James spent the early years of his life on a country estate, Glenlair, where he received schooling only from his mother. She died when he was eight, which affected him deeply, and two years later he was sent to Edinburgh Academy, and then at 16 to Edinburgh University. He excelled in the study of philosophy, religion, and mathematics. He went on to obtain degrees at Edinburgh and Cambridge Universities and was a professor at various times in Mathematics and in Physics. He married Katherine Mary Dewar in 1858, the daughter of the principal of his college at Edinburgh, who was seven years older and helped in his lab, in particular working on experiments in viscosity.

By 1858 Maxwell was already famous for his theory of Saturn's rings, which had been discovered by Huygens as we related in Chapter 2. Maxwell showed the rings are not a liquid or gas, but instead most likely consist of a huge number of independent, rock-like particles in relatively stable orbits about Saturn, a theory finally verified by the 1980s' Voyager flybys. Later he made many even more important, fundamental advances. He developed the kinetic theory of gases which explained properties of gases by the motions of their molecules. He explored the concept

Fig. 10. James Clerk Maxwell as a youth and as a famous scientist with his wife and dog.

of entropy in statistical mechanics and clarified it by inventing the idea of Maxwell's Demon, a wily creature who seemingly violates the law of entropy by opening or shutting a trap door connecting two containers depending upon the speed of an approaching molecule.[11] But his most important contribution was in the field of electromagnetism.

Maxwell decided he would read Faraday's *Researches* before tackling the great questions about electromagnetism. He became convinced the correct approach is Faraday's; namely, that electromagnetic effects are propagated by fields from one place to another rather than simply acting at a distance. If a charge suddenly moves, the resulting effect on another charge is not instantaneous but is delayed until the disturbance created by the moving charge propagates to the location of the second charge. What is the speed of propagation? As we have seen, disturbances can move on telegraph wires at the speed of light. What is the speed if there is only free space between the charges? It seemed to Maxwell almost certain that it was the speed of light also.

While attempting to answer that question, Maxwell saw that Ampere's Law (see Fig. 6 and Eqs. (4) and (5)) had to be modified to include what happens when a steady current charges a capacitor (a device consisting of two conductors very close together) as in Fig. 11. He noticed that the current flowing into the capacitor causes charge to accumulate on the conductors,

---

[11] The solution is to include the entropy of Maxwell's Demon itself.

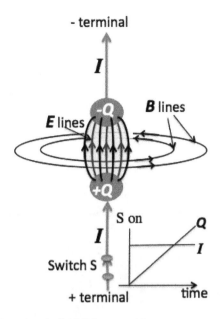

Fig. 11. An increasing electric field E is created between two conducting balls as they build up a charge Q from the current I that flows when the switch S is on. The *changing E* produces a field B as indicated in the loops just as the current I does in Ampere's Law as shown in Fig. 6.

producing a changing electric field E between them as shown in the figure. Maxwell hypothesized that this changing E produces a field B around the loop just as a current does, hence the name *displacement current* we mentioned above, and modified Ampere's Law to include both effects:

$$Loop\ B = \mu_0\ I + \mu_0\varepsilon_0\ \Delta\Phi_E/\Delta t \qquad (7)$$

where *Loop B* has the same meaning as in Eq. (5) and $\Delta\Phi_E/\Delta t$ *is* the time rate of change of the electric flux, $\Phi_E = EA$, through any area A bounded by the loop. Using Eq. (7) we obtain the correct result for B whatever the loop and area A bounded by it. Equation (7) is Maxwell's 4th equation, which has been verified by experiment. When combined with his other equations, it predicts EM waves traveling at the velocity of light as we will now discuss.

Maxwell recognized that Eq. (7) was the missing link. When he combined it with his other equations, Maxwell was able to show that an

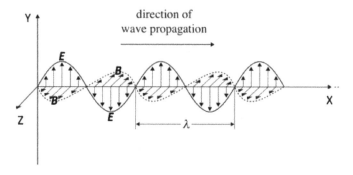

Fig. 12.   An EM wave of wavelength λ propagating in the X direction, showing the *E* and *B* fields just on the X axis where $z = y = 0$. The same pattern of *E* and *B* occurs for all axes parallel to X, e.g. for all values of z and y in the region where the wave exists.

Fig. 13.   Heinrich Hertz.

oscillating charge would produce at great distances an EM wave like that shown in Fig. 12. In this wave, *E* and *B* are sinusoidal fields that are perpendicular to each other and are each perpendicular to the direction of propagation. The value of the fields at any point gives the force that would act on a charge or current at that point. For clarity, the figure shows the values of the fields only for points on the X axis, but the same pattern of fields exists on all axes parallel to X within the lateral extent of the wave. Thus every point in a YZ plane has the same values of *E* and *B*, which change with time as the wave propagates past them at some speed v, which we can prove equals c, the velocity of light. We will only outline the proof. If we apply Faraday's Law (*Loop E* $= -\Delta\Phi_B/\Delta t$) to some loop in the XY plane it is not too complicated to show that $E = vB$. In like manner, if we use a loop in the XZ plane and apply Eq. 7 when $I = 0$ (which is *Loop B = $\mu_0\varepsilon_0\Delta\Phi_E/\Delta t$*) we would obtain $B = v\mu_0\varepsilon_0 E$. Combining the two underlined equations we obtain $v = c$ and $E = cB$ where we have used $c^2 = 1/\mu_0\varepsilon_0$ found earlier. So we find that EM disturbances travel at $c = 3 \times 10^8$ meters/sec, the speed of light but determined entirely by electric and magnetic measurements. As a bonus, we find that in an EM wave, E is always c times B in the units we use.

How are the fields E and B related to the energy carried by an EM wave? J. H. Poynting was able to show that the time average of EB/μ_0 over a complete period of the wave equals the average energy per unit time per unit area transported by any EM wave, including a light beam. So, measuring the power per unit area delivered by a light beam allows one to determine the size of the E and B fields in the beam. A very important problem remained to be solved, namely how light as an EM wave is emitted and absorbed by atoms. This problem was not solved until the 20th century, as we will learn in Chap. 10.

Maxwell's equations clearly explain light as an EM wave, but they also predict EM waves of a much lower frequency that could in principle be generated by oscillating charges and currents in the laboratory. No one could think of a way of verifying this prediction until Heinrich Hertz came up with an idea in 1885. Hertz was born in 1857 to a wealthy and well-known family in Hamburg, Germany. He attended a small private school for 10 years where he excelled in mathematics, science, and foreign languages, and also in practical subjects such as woodworking and machine shop. After high school and a year of voluntary military service, he became a student of the great physicist Hermann von Helmholtz at Berlin. He received his doctorate, and then became a professor at the University of Karlsruhe. There he met and married Elizabeth Doll, the daughter of one of his fellow professors. At Karlsruhe he began his search for EM waves generated by electric circuits. This problem had been suggested to him earlier by Helmholtz in graduate school, but he did not feel prepared for it until he had studied Maxwell's equations thoroughly.

Then Hertz had a brilliant idea. He of course was familiar with the induction coil we mentioned earlier in footnote 9 in connection with Faraday's Law. Hertz connected the two secondary (high voltage) leads of the coil to two electrodes separated by an air gap, with each electrode

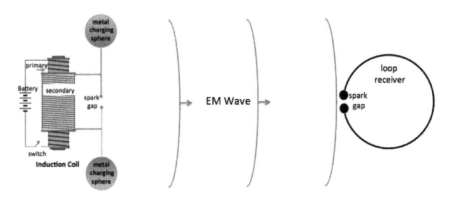

Fig. 14.   Experiment by Hertz in which radio waves were generated and detected for the first time.

attached to a large conducting sphere, as shown in Fig. 14. When the induction coil is switched on, high voltage builds up across the gap and charges up the conducting spheres, continuing until the break-down voltage of the air is reached in the gap. At this point, the gap becomes conducting and discharges the spheres, creating a large arc or spark across the gap. The discharge will be oscillatory (like a plucked guitar string), and the frequency of the oscillation can be calculated from the size of the spheres, the length of the wire connecting them to the gap, and other circuit parameters. Using Maxwell's equations, Hertz could then calculate the wave that would be broadcast by the current and voltage at this frequency. This wave could be detected some distance away by the spark it excited across a gap in a circuit tuned to the same frequency and located as in Fig. 14. This response is somewhat similar to that of a tuning fork set at the right frequency when it vibrates sympathetically with a loud tone generated nearby.

This was not only a brilliant idea, but Hertz also knew how to build the apparatus to test it! The frequency for his apparatus was about 450 million cycles per sec, or 450 MHz, where Hz stands for Hertz, the official name for a cycle per second. In a remarkable series of papers he sent to Helmholtz, Hertz verified that Maxwell's Equations correctly described the EM wave he generated and detected. From the measured wavelength λ and frequency ν, he further verified that $\lambda\nu = c$, the speed of light. As fate would have it, however, while doing this experiment Hertz noted the detector sparks were enhanced by UV light. This observation, as we will see in the next chapter, helped usher in the idea of the *particle nature* of light, ironically just as he was verifying the existence of EM *waves*.

Only a few years after his famous discovery, Hertz died on New Year's Day in 1894 of a bone malignancy at the young age of 37. His research ushered in the modern communication age, including wireless telegraphy, radio, TV, etc., all based on EM waves he was the first to generate.

Hertz's research also got people thinking about how charges bound inside atoms and molecules could generate and interact with light. These charges were soon identified as electrons. Physicists hypothesized that these electrons could emit and absorb light by vibrating about their equilibrium positions, since moving charges emitted and absorbed radio waves in Hertz's experiments. Of course, the bound electrons would have to vibrate at the very high frequencies of light. Although this idea was greatly modified by quantum theory in the 20th century, it still sufficed to explain refraction and dispersion of light by transparent materials, so that people realized they were on the right track. The Dutch physicist H. A. Lorentz is credited with making the greatest progress with this idea in the last years of the 19th century, which is fitting since his fellow countryman, Snell, found the right laws of refraction two and a half centuries earlier. Refraction is especially relevant for us because it is the problem inherited from antiquity whose solution by Snell and Descartes is the story with which we started the book, back in Chapter 1.

We can give only a brief summary of Lorentz's treatment. When a visible light beam enters a transparent medium, very little light is absorbed because the natural oscillation frequency of the bound electrons is far higher than the frequency of visible light. In this case the oscillating electric field of the light sets the bound electrons into oscillation at the same frequency as the light, causing them to radiate more light by Maxwell's equations. In the direction of the initial beam (what we call the forward direction) the phase of this "scattered" light lags behind the phase of the initial beam, pulling back the net phase of the combined beam and slowing it down in the medium. This explains refraction, as in Fig. 4 and Eq. 1 in Chapter 2. Dispersion is also explained by this model, for if we change the incident light from red to blue, i.e. from lower to higher frequency, the electrons are driven at a frequency closer to their resonant frequency and oscillate with a larger amplitude. Thus blue light produces a larger forward scattered beam than red light, which pulls the resultant wave back further and slows it down more than does red light, leading to a greater refractive index as we go from red to blue. Miraculously, this entire discussion remains valid when the electrons are described by the quantum theory even though much else is changed drastically by that theory, as we will see in the next chapter.

So Lorentz's theory explains why light travels slower in most transparent materials, a fact we expected from Huygens' wave theory of refraction back in Chapter 2. And we get a huge bonus: an explanation of dispersion, namely why the index changes with color and thus a white beam splits into colors when sent through a prism as studied by Newton in Chapter 3.[12]

So, what about the luminiferous aether? That issue was finally settled shortly after the end of the 19th century, although explaining light as an electromagnetic phenomenon did not at first dispel the idea of the aether. In fact, both Maxwell and Hertz were ardent believers in it. The statement by Maxwell in the epigraph introducing this chapter presupposes this belief. Electromagnetic actions were held to be propagated as distortions of some kind of medium, i.e. the aether. However, problems began to arise. The Michelson–Morley experiment described in the previous chapter, in which no motion of the Earth through the supposed aether could be detected by its effect on the speed of light, certainly posed a major problem for aether proponents. Then there was the idea that the electromagnetic field did not really need an aether if one accepted the field as being a real primary entity in the same sense that a material object is a real entity; in fact, once the electric nature of matter was understood, in which electric and magnetic fields exist inside atoms, EM fields seemed simpler than ordinary matter which is a complex combination of fields and material particles. Why then would one insist on explaining a simpler phenomenon by ideas derived from the properties of a more complicated one? At this point, the whole idea of needing a medium to vibrate and produce EM

---

[12] Lorentz's theory had other successes, including his explanation of the splitting of spectral lines in a magnetic field discovered by Pieter Zeeman in 1896. From the size of the splitting, Lorentz and Zeeman calculated the ratio of charge to mass of an electron, e/m. The same value of e/m, published a month later than the Zeeman effect in the same journal, was found in laboratory experiments on a beam of electrons by J. J. Thomson, whom we'll meet again in Ch. 10. This proved that the electrons seen in the lab were the same particles as those accounting for radiation in atoms.

waves just died away. *E* and *B* fields were simple entities that do not need distortions of another substance to exist. In fact, an EM wave in free space, such as a beam of light, is one of the simplest objects we know about, far simpler than an atom![13] There are also arguments from the Theory of Relativity about the aether which played an important historical role. In the end, the notion of an aether stopped being mainstream, although since it is difficult to *disprove* the idea (since one can endow the aether with any property one needs) it still pops up from time to time as an outlier theory.

In this chapter we have reviewed the huge field of electromagnetism, and indicated how Maxwell used it to predict the existence of EM waves traveling at the velocity of light, a prediction that was verified by Hertz for radio waves in the 1880s. From that point on, physicists knew that light is an EM wave, and all the properties of light waves we have learned about in the first 8 chapters of this book are explained by Maxwell's equations of electromagnetism. *Thus two vast fields, optics and electromagnetism, were unified into one.* It was a great triumph, but there was much more to learn about the nature of light, and hence electromagnetism, as we will see in Chapter 10.

---

[13]The question of whether an elementary particle, such as an electron with its associated electric and magnetic fields, is as simple as a pure EM wave is not one to be tackled in this book!

Chapter 10

The Entrance of the Photon and the Birth of Quantum Theory

"He was, by nature, a conservative mind; he had nothing of the revolutionary and was thoroughly skeptical about speculations. Yet his belief in the compelling force of logical reasoning from facts was so strong that he did not flinch from announcing the most revolutionary idea which ever has shaken physics."

Max Born (great physicist in his own right), commenting on Max Planck and his concept of the photon.

Light is a wave. That has been the message so far in this book. All phenomena we have studied, including bending of light by refraction, interference of light from thin films and multiple apertures, diffraction of light by small objects, precision measurements with the interferometer using light from spectral lines, and connecting light with electromagnetism were elucidated by the wave theory of light. However, near the beginning of the 20th century, the work of many scientists, the earliest being Max Planck and Albert Einstein, revealed that light is a particle as well as a wave. These particles are *not* the particles Newton invoked to explain the familiar properties of light. The wave theory explains all of *them*. What kind of particle are they then and what do they explain that the wave theory can't? They are called light quanta or photons and for a light wave of frequency ν they have an energy E given by

$$E = h\nu \tag{1}$$

where h is Planck's constant, a universal constant of nature. We will learn a bit more about the concept of energy and then we will see that photons explain a very common observation, the light radiated by hot objects. They also explain the photoelectric effect, the emission of electrons when light shines on metal surfaces. The wave theory fails in both cases.

There was more to come. Following the work of Planck and Einstein, Niels Bohr postulated that atoms have only discrete energy levels and that a photon is emitted or absorbed when an atom makes a transition between two of these levels. The photon energy $h\nu$ equals the energy difference between the two levels, which finally solved the longstanding problem of why the Fraunhofer lines of Chapter 8 have a specific frequency, a problem the wave theory also failed to explain. Bohr derived a theoretical formula for the energy levels of the hydrogen atom, which was *the same formula* as the empirical Balmer/Rydberg formula in Chapter 8 for the spectral lines of hydrogen. Bohr's formula gave the exact value of the empirical Rydberg constant, another great triumph. Bohr's theory launched a whirlwind of creative activity in the 1920s out of which the general theory of quantum mechanics was born. We will touch on just a small part of this theory to see that material particles, such as electrons, behave like both a wave and a particle just as photons do. We will end with the part of quantum theory that resolves the apparent contradiction when light (or the electron) is both a wave and a particle.

To get started, we first inquire how the energy carried by a light beam is measured. One method was perfected in the late 19th century, namely focusing the beam onto an object called a *black body*, i.e. an object that absorbs all the light incident on it which we will say more about shortly. The energy absorbed shows up as a rise in temperature of the body, the brighter the light and the longer the time it is on, the higher the temperature. So one form of energy, light, is converted into another form, heat. This is an example of the *conservation of energy*. To measure how much energy the beam carries, we can measure separately how much heat energy in joules[1] is required to raise the temperature of the black body the same amount as absorbing the light does. Then, knowing the cross-sectional area of the light beam and the time it is on, we can determine the energy per second per unit area of the light beam. This is the *intensity I* of the beam, expressed often as watts per cm^2 where 1 watt = 1 joule per sec. (As a

---

[1] One joule of energy will heat 0.238 g of water by 1°C.

Fig. 1.   Common sources of heat radiation.

reminder, the intensity was given in terms of the EM fields of the beam in Chapter 9.) If the beam has a spread in wavelength $\Delta\lambda$, then $I/\Delta\lambda$ is the intensity per unit wavelength.

Now we can turn to the problem that Planck and others encountered. The wave theory of light, successful as it was, could not explain the intensity of radiation from a heated body. As an object such as a heater wire or an electric heating element on a stove is heated up, we see it first glow dull red, then shift to orange if it becomes hotter; an incandescent light bulb or candle flame can become hotter still and then glows white. These observations are illustrated in Fig. 1.

The objects of Fig. 1 emit most of their radiation in the infrared; even for the tungsten light bulb, it is just the tail of emission in the visible that we see. The complete picture is given in Fig. 2, in which the light output intensity per unit wavelength for a *black body* is plotted as a function of wavelength λ for several temperatures.

Black Body Radiation

Although most objects radiate in a similar way to what is indicated in Fig. 2, the actual radiation curves shown in the figure are not from any old object but instead from a reproducible object called a black body. Just as in common parlance, a black body is black because it absorbs all radiation incident on it. It may seem a little surprising that such a body is a good radiator too. For precise measurements, a black body is well-approximated by a large

---

(*Continued*)

(Continued)

cavity with only a small hole connecting it to the outside. The hole is the entrance area of the black body. Any radiation incident on the hole from outside rarely escapes back out because the hole is small; at low temperature the cavity so illuminated appears black to the outside when seen through the hole. This property we have already used to measure the energy of an incident light beam. Now, if the cavity wall is heated up it glows through the hole, and that is the black body radiation that is plotted in Fig. 2. This spectrum could be measured very accurately in the last half of the 19th century.

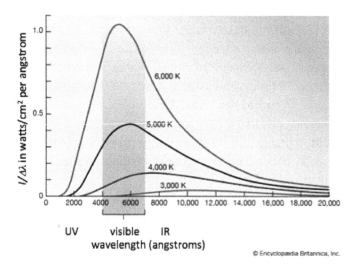

© Encyclopædia Britannica, Inc.

Fig. 2. Blackbody spectrum for several temperatures given in degrees Kelvin (K). Temperature in K equals temperature in Celsius plus 273.

The correct theory of radiation sought by Planck and others should be able to predict this behavior. The wave theory could not come close. In fact, the electromagnetic wave theory predicted steadily increasing radiation as the wavelength gets shorter, becoming infinite for all temperatures!

This state of affairs was very unsatisfactory to say the least. Eventually, Max Planck, a renowned physicist not given to wild, revolutionary ideas, came up with just that, the revolutionary idea that light is emitted and

Fig. 3. Max Planck (1858–1947) came from an intellectual family of three generations on both sides. In school he excelled in physics and music (playing piano, organ, and cello and composing songs and operas) but pursued physics in graduate school in Munich and Berlin under Helmholtz and Kirchoff. He was Professor of Physics at Berlin for decades and the acknowledged leader of German physics for the turbulent years of the two world wars. He became a close friend of Einstein and a fellow musician, as well as a firm backer of his Relativity Theory when it was first proposed.

absorbed in units of energy $h\nu$, where ν is the frequency of the light and h is called *Planck's Constant*. His idea is expressed in Eq. (1) above. He reasoned that if light were required to radiate in ever larger units $h\nu$ as the frequency ν gets higher, then these units would get ever harder to create as ν increases, and thus the blackbody curve should turn over and start decreasing at higher frequencies, just as it does in Fig. 2. Planck used the theory of statistical mechanics to make this crude, tentative idea very precise, and the resulting formula is called Planck's Law, with one free parameter, h, which Planck determined by fitting his curve to the known blackbody spectrum of Fig. 2, yielding the value $h = 6.602 \times 10^{-30}$ Joules per Hertz. With this value, the fit of Planck's theoretical curves to those in Fig. 2 is nearly perfect. The extremely small value of h in ordinary units meant that light of even very small intensity would involve a huge number of photons per second being emitted or absorbed by areas of ordinary size, and so light would seem to be continuous in ordinary experiments. But the effect of these photons had enormous repercussions beyond blackbody radiation, and led to the quantum theory of light and eventually even to the quantum theory of material particles. Planck received the 1919 Nobel Prize for this breakthrough.

One problem was how to reconcile the mountain of evidence that light is a continuous wave with this strange idea that it comes in irreducible quanta of energy $h\nu$. That problem didn't get resolved until many years later, as we will see at the end of this chapter. Meanwhile, Planck himself did not take the full leap of assuming light beams intrinsically consist of these quanta, but hedged, thinking that light was only emitted or absorbed in such units. Five years later, Einstein did take the full leap when he explained the photoelectric effect, which we take up next.

There had been a hint of something amiss as far back as 1887, just as the wave theory of light was reaching its triumphal peak with the detection of EM waves by Hertz discussed in Chapter 9. There was what appeared to be only a minor unexplained observation in Hertz's experiments. As described in Chapter 9, Hertz detected an EM wave by the spark it induced across the gap of a receiver coil. He happened to have this receiver gap facing the spark gap that generated the waves. In hopes of seeing the detecting spark better, he blocked out ambient light by enclosing the coil in a box with a glass window facing the source circuit with its own spark gap. Instead, the detected spark length got shorter. He quickly realized what the problem was. The box had that glass window facing the source spark gap. When he removed the glass, the detected spark was restored to its original length. When the glass was replaced with a window of quartz, which transmits UV light, the length of the spark was also restored. So the UV light from the source spark was somehow helping make the detection spark longer. Hertz went on with his EM experiments, but wrote up his strange observation. Others followed up on this observation and showed conclusively that shining UV light on the detection gap enhanced the spark by liberating negatively charged particles from the metal surface, as illustrated in Fig. 4(a). This phenomenon became known as the *photoelectric effect*, and the liberated particles were shown to be electrons, called photoelectrons when produced in this way.

The systematic study of the photoelectric effect was then undertaken using an evacuated glass tube like that shown in Fig. 4(b) with a window for admitting a UV light beam onto a metal surface to create photoelectrons. These electrons then landed on a collection plate producing a

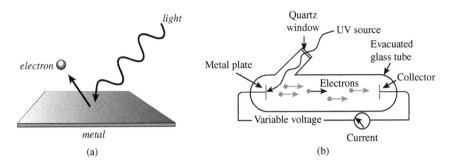

Fig. 4. The photoelectric effect.

photocurrent that could be measured by an ammeter and was found to be directly proportional to the intensity of the light, as shown in Fig. 5(a). We will explain this proportionality later.

Earlier we saw how light energy could be absorbed and converted to heat. Now we see that in the photoelectric effect, at least some light energy can be converted into the energy of photoelectrons. So we ask: how much energy do the ejected electrons acquire from the light? They certainly acquire their energy of motion, called their kinetic energy, *KE*, that they have just as they leave the emitter plate. They also acquire the energy to free them from the metal; this is called the work function of the metal *W*. So the energy supplied by the light to each photoelectron is

$$\text{Energy given to each photoelectron} = KE + W \qquad (2)$$

The initial *KE* could be determined by using the voltage V of the collector plate relative to the emitter plate as we will now explain. Since the emitted electrons have a negative charge, if *V* is positive, the electric field between the plates accelerates the emitted electrons, but if *V* is negative, the field decelerates them. The energy of the ejected electrons was found by first setting the sign of *V* to retard the photoelectrons and decrease their *KE*, and then finding the magnitude of the retarding voltage that prevented them from reaching the collector at all. When the retarding voltage reached the value that stopped all the electrons, called the *stopping potential*, that value would correspond to the maximum kinetic

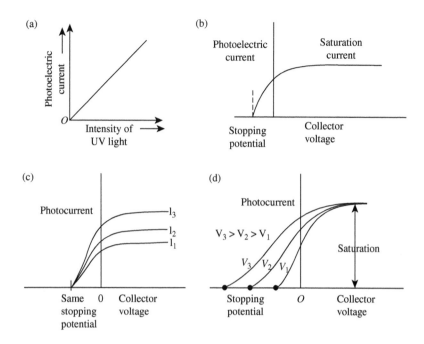

Fig. 5.   Properties of the photoelectric effect.

energy that electrons had acquired from the light. So $KE = eV_{stop}$, where e is the magnitude of the charge on the electron. This behavior is shown in Fig. 5(b).

Then in 1902 Philipp Lenard, Hertz's former assistant, made an observation that was completely at odds with the wave theory of light. When he increased the intensity I of the light, even by seventy fold, the stopping potential was unchanged. In other words, the electrons acquired the *same maximum kinetic energy from the light regardless of its intensity* as shown in Fig. 5(c). Just more photoelectrons were ejected with increase in light intensity, as shown by the photocurrent in the same figure and in Fig. 5(a).

So what *does* determine the maximum kinetic energy? Recall that Hertz had observed the photoelectric effect when only UV light shone on the electrodes. So it is the wavelength (or equivalently, the frequency) of the incident light that determines how much energy individual photoelectrons absorb, not light intensity. In Fig. 5(d), we see that the higher the frequency of the light ν, the larger the retarding potential required to stop the photoelectrons; in other words, the greater the initial photoelectron

Fig. 6. Albert Einstein (1879–1955) and Isaac Newton are often called the two greatest physicists of all time. Einstein was born in Ulm, Germany. While working at a Swiss patent office in 1905 he published three landmark papers on (1) the photoelectric effect as discussed in the text, (2) the Special Theory of Relativity, and (3) Brownian motion, the result of small particles buffeted about by molecular collisions. In 1916, he developed the General Theory of Relativity to explain gravity and it is still revealing new properties of the universe. He was a professor at Berlin until he came to Princeton in 1933.

energy. None of this had an easy explanation in the wave theory in which the intensity of the wave, not its frequency, should determine how much energy the electron absorbs. If you have guessed that this behavior has to do with Eq. (1) above, you are right.

In 1905, Albert Einstein made that guess and found the solution to this dilemma. He started with Planck's hypothesis that the emission and absorption of light takes place in discrete units or quanta equal to $h\nu$, which we saw solved the blackbody radiation problem. Then he proposed that *light itself is quantized in discrete units hv, called photons,* and that is the reason that emission and absorption could take place only in units of these photons. With this assumption, which is Eq. (1) above, he solved all the problems posed by the photoelectric effect. The emitted energy of each photoelectron and hence the stopping potential should rise directly with the energy $h\nu$ of the photon it absorbs, thus explaining Fig. 5(d). If ν does not change, the stopping potential would not change either as in Fig. 5(c), but the number of photoelectrons would increase with light intensity, as shown in the same figure, since higher intensity means more photons to produce photoemission. In other words, the energy supplied by the light to each photoelectron is the energy $h\nu$ of a photon, so Eq. (2) becomes:

$$h\nu = KE + W \tag{3}$$

For this work, Einstein was awarded the Nobel Prize for physics, but not until 1921. The delay was partly because the experimental results were only approximate in 1905. Precise results were not easy to obtain because exposure of the tube in Fig. 4 to the air, before being sealed off, would change the work function of the metal surface rapidly — often within seconds — due to oxidation and other contamination. Finally, in 1914 the American physicist R. A. Millikan obtained accurate data by the herculean effort of building a vacuum chamber around a lathe that held a turning sample of the metal whose surface was continually scraped clean. Millikan proudly described this tour de force as "constructing a machine shop in vacuo."[2] He found that with his data the slope of the curves giving the stopping potential vs. light frequency gave *the same value of h* as Planck had found from the blackbody data — just as Einstein had predicted.

Brief Side Note on the Compton Effect

Confirmation of the photon nature of X-rays, which have a much shorter wavelength than even UV light, came from an impressive experiment by Arthur Holly Compton in 1923 in which he observed the scattering of a beam of X-rays by the electrons in a piece of Carbon. Maxwell had shown many years earlier that an EM wave carrying energy E has a momentum of E/c, where c is the speed of light. Compton reasoned that an X-ray photon of energy $E = h\nu$ would have a momentum $E/c = h\nu/c = h/\lambda$, and the scattering in his experiment could be described accurately by the laws of conservation of energy and momentum of two colliding particles — the photon and the electron. The electron would recoil when struck by the photon, leaving the scattered photon with correspondingly less energy and a longer measured wavelength λ'. The energy lost by each photon $hc/\lambda - hc/\lambda'$ could be calculated for each angle of scattering, and agreed precisely with Compton's measurements.

---

[2] Millikan was awarded the Nobel Prize in 1923 for these photoelectric measurements and for his measurement of the electron charge.

Fig. 7.   Ernest Rutherford (1871–1937).

Don't be misled by the simplicity of this result for the photoelectric effect or the Compton effect. (That is, for those readers who do think the result is simple!) The wave nature of light still lurks there. For example, the energy $h\nu$ of the light *particle* is expressed in terms of the frequency ν of the *wave*. Furthermore, the wave nature of light determines where the photons will be found. For example, if a single beam of light hits a cathode, photoelectrons will be produced; however, if a second beam from the same source but coming from a different direction also hits the cathode and forms an interference minimum with the first one at the cathode so that they cancel each other out, i.e. a wave phenomenon, no photoelectrons will be ejected. So light has wave properties and particle properties. We will study this dilemma further at the end of this chapter.

Let us now turn finally to the question that we left hanging in Chapter 8. How is light emitted and absorbed by atoms? We know the answer must be consistent with the sharp lines, the Fraunhofer lines, that we see in spectra. We will see that the discovery of photons provides the key to answering this question. First, we will sketch the prevailing view of the structure of atoms around 1900. It was known that neutral atoms consist of negatively charged electrons that are responsible for emission and absorption of light, and a positive charge about which little was known except its charge balanced the charge of the electrons. The most popular model that incorporated these facts was developed by J. J. Thomson and was called the pumpkin model of the atom (or sometimes the plum pudding model) in which the negative electrons were like seeds held inside the soft positively charged pumpkin. The electrons could vibrate about their equilibrium position inside the pumpkin, at the frequency of absorption or emission of radiation. In other words, the electrons would act like little radio wave antennae, but operating at the much higher frequencies of light.

Then this model was completely overthrown by the classic scattering experiments of Ernest Rutherford. These experiments grew out of the discovery of radioactivity by Henri Becquerel in 1896 and the identification of radium as one of the prime sources of this mysterious activity by Marie and Pierre Curie two years later, for which all three were awarded the Nobel Prize for physics in 1903. Marie Curie was then awarded a second Nobel Prize (in chemistry this time) in 1911 for showing that Radium is an element, putting her in a select group who have won the prize twice. Radium was the source of radioactive particles later used by Rutherford in his experiments.

Rutherford was born in Brightwater, New Zealand, in 1871 to British émigré parents. He won a fellowship to the Cavendish Laboratory to work with J. J. Thomson and after receiving his doctorate he was offered a position at McGill University in Canada, where he commenced research on radioactive sources. He discovered the three principal emissions of such sources, what he called α, β, and γ rays, and which he later identified as He nuclei (α), electrons or positrons (β), and very high energy photons (γ), respectively. He made many more advances in radioactivity and won the 1908 Nobel Prize in Chemistry for this work. He then went to the University of Manchester in England and started the work for which he became even more famous, the scattering of α particles by a gold foil to probe the internal structure of the gold atoms in the foil.

The thin foil of gold was placed near a radioactive source of energetic α particles as shown in Fig. 8(a), and the particles that were scattered by the atoms in the foil were detected at various scattering angles by a sensitive screen. If the positive and negative charges were spread throughout the atom, as envisaged in the Thomson pumpkin atom, then it was thought that such fast α particles would go through each atom with only a minor

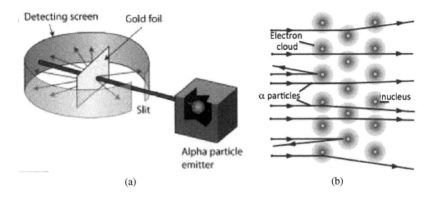

Fig. 8. Rutherford's experiment. (a) Schematic of the experiment. (b) A closer look at what is happening at the gold foil.

deflection. Instead, Geiger and Marsden, Rutherford's assistants who actually carried out the experiment, saw that a few were scattered through large angles, and some came almost directly back, as shown in Fig. 8(b). They were able to show that such large angle scattering could take place only when the α particles came very close to a highly charged and massive body, which must be the nucleus of a gold atom with all its charge Q and mass concentrated in a tiny region at the center of the atom. In fact, the scattering pattern could be explained by the $Q/4\pi\varepsilon_0 r^2$ Coulomb field we mentioned early in Chapter 9, produced in this case by the small charged nucleus at the center of each atom. Therefore, the much larger volume of the atom must be occupied by the atomic electrons. This result led to the so-called planetary model of the atom, where all the positive charge is concentrated in the nucleus, a small region at the center of the atom about which the negative electrons move in orbits, as pictured in Fig. 9. This picture is roughly the one we use today.

This was the situation when the young Danish physicist Niels Bohr arrived in Rutherford's lab in 1912. Bohr, who was to become one of the most important theorists of the first half of the 20th century, was born in Copenhagen, Denmark, in 1885.[3] It did not take long for Bohr to be con-

---

[3] He and his brother, Harald, were outstanding soccer players. Harald played on the Danish team in the 1908 Olympics and earned a silver medal. Harald became an eminent mathe-

Fig. 9.   The planetary model of the atom.

Fig. 10.   Niels Bohr.

vinced by the Rutherford planetary view of the internal structure of atoms, and when he went back to Denmark, he began to use this view to solve the problem of atomic spectra. He knew that the simplest atom, hydrogen, consists of a negatively charged particle, the electron, in orbit about a much heavier positively charged particle, what we now call the proton. He also was aware that the frequency of each line of the spectrum of hydrogen is equal to the difference of two terms, a formula given in Chapter 8 (Eq. (3)); this formula had lain a mystery of simplicity for 25 years.

matician and Professor at Copenhagen.

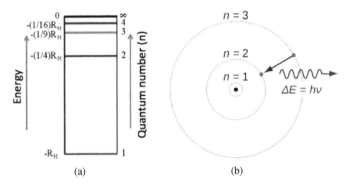

Fig. 11. (a) The energy levels of the hydrogen atom. (b) Bohr's concept of the radiative transition between two states made by emission of a photon of energy $h\nu$ equal to the energy difference between the two states.

Bohr then suggested that each term corresponds to an energy level of an electron orbit in the atom, and the difference between terms, the actual frequency of spectral lines, corresponds to a transition between orbits, as shown in Fig. 11 for hydrogen. The terms thus represent energy levels in the atom, and the energy of the photon released or absorbed in the transition, $h\nu$, is equal to the energy difference between two levels.[4] *Note the central role of the photon energy $h\nu$.*

Bohr endeavored to find an expression for these energy levels in the hydrogen atom based on solving for the orbit of an electron moving in the known Coulomb electric field of the proton. (To avoid getting sidetracked, we give here just a very brief outline of how he did it.) Clearly, only certain special orbits are allowed, a different orbit for each energy level. It can be seen from Fig. 11 that these levels involve an integer n, the same integer as appeared in Eq. (3) of Chapter 8. Bohr called n a quantum number with $n = 1, 2, 3$, etc., and found the quantization condition for an electron of momentum p (defined as the mass m times the velocity v, or mv) moving in a circular orbit of radius r:

---

[4]Bohr actually applied this statement to all atoms. He knew that other atoms besides hydrogen had spectral lines, which also could be expressed as the difference of two terms; they were just more complicated than for hydrogen.

$$pr = nh/2\pi, \qquad\qquad (4)$$

where n is the same integer as in Fig. 11. (In most discussions of the Bohr theory, Eq. (4) is just postulated to be true.) The details are complicated, but Eq. (4) leads to the result for the energy of an electron in the H atom for each value of n: $E_n = -R_H hc/n^2$ with $R_H = 4\pi^2 me^4/h^3 c$, where e is the charge of the electron. Putting in the known values of the quantities yields $R_H = 1.0968 \times 10^{-7}$ per meter, the same value as found years earlier experimentally for the spectrum of atomic hydrogen (see Eq. (3) in Chapter 8), and occurs in the formula along with the same dependence on n! No one could argue with such quantitative accuracy, and Bohr won the Nobel Prize in Physics in 1922.

The Bohr theory of the H atom clearly was on the right track, with its great quantitative success in predicting the energy levels and spectral lines. But it got stalled immediately when it was applied to a more complicated system. Even the Helium atom with just two electrons was not solvable. It turned out that the correct theory to describe the quantum world was much stranger yet. With the Bohr model, however, optical transitions and other atomic processes could at least be visualized, though not calculated accurately, and the model is still used today for pictorial purposes.

We will illustrate with one example, but it is certainly a major one: the laser. Laser stands for Light Amplification by Stimulated Emission of Radiation, and we will first describe what Stimulated Emission is. Let us consider two atomic energy levels shown in Fig. 12 between which an optical transition of frequency ν can take place, just as in the Bohr model. If the atom is in the lower of these levels, and light of frequency ν shines on it, the atom can absorb a photon and make a transition to the upper level, as shown in Fig. 12(a); a number of such atoms could produce a dark Fraunhofer line. If instead the atom is in the upper state, it can spontaneously emit a photon of frequency ν, in a range of directions as shown in Fig. 12(b). A number of such atoms in the upper state could produce a bright Fraunhofer line. There is a third possibility. The atom is in the upper state, and a photon of frequency ν comes along and stimulates it to emit another photon before the atom can spontaneously emit, as shown in Fig. 12(c).

The remarkable feature of stimulated emission is that the emitted photon has identical properties to those of the stimulating photon; they

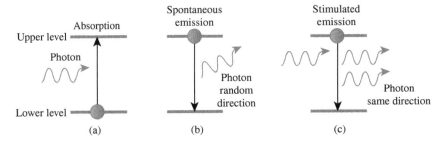

Fig. 12. Three ways in which photons of energy equal to the energy difference between two atomic levels are absorbed or emitted. (a) Absorption of an incident photon by the atom when it initially is in the lower level. (b) Spontaneous emission when the atom is initially in the upper level. (c) Stimulated emission of a second photon identical to an incident photon when the atom is in the upper state. This process is central to a laser.

have exactly the same frequency and phase, so that there are no two identical photons. Furthermore, these two photons have a higher probability than a single photon of stimulating another atom that is in the upper state to emit another photon. In this way, one photon of the right frequency can create a large number of photons of exactly the same phase and frequency if there are enough atoms in the upper state (and only a few atoms in the lower state to absorb the photons). This is the Light Amplification part of the Laser. A medium that has a large excess of these upper state atoms is called a gain medium. A medium with enough gain will produce laser oscillation because spontaneous emission always occurs to produce a few photons to get the process going. These three mechanisms of radiation were first analyzed by Einstein in 1917 with Bohr's picture of atomic energy levels still the prevailing view. It was not until 1960, however, that lasers (and a few years earlier, masers using Microwave Amplification) were conceived and developed.

The HeNe laser, as it is called, provides a good illustration of how these ideas work out in practice; namely how a *population inversion* is established, the term used when the upper level has a larger population than the lower level, and how the photons created by stimulated emission from the upper to the lower level can build up to create a laser beam. A mixture of helium and neon gases is contained in a tube with attached electrodes maintaining an electric discharge inside it, as depicted in Fig. 13.

The laser operates on neon energy levels as shown in Fig. 13(a), where 5s is the upper level and 3p is the lower level of the laser transition

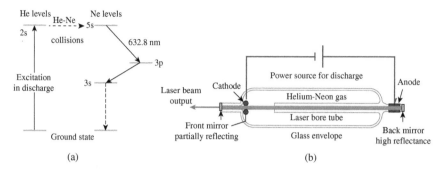

Fig. 13. (a) The energy levels of He and Ne germane to the operation of the HeNe laser showing the 632.8 nm laser transition. (b) The apparatus used in a HeNe laser.

at the wavelength 632.8 nm.[5] A population inversion between these levels is established because the 3p state is depleted by radiation to lower states while the 5s state is constantly populated by collisions of ground state neon atoms with excited state He atoms, which have almost the same energy as the 5s neon atoms and thus can readily transfer their energy over to the neon. These excited He atoms are produced in the discharge. The laser then operates on the 5s–3p transition, with the 632.8 nm photons created by stimulated emission reflected back and forth through the medium by the mirrors at each end of the tube shown in Fig. 13(b), which enables each photon to make many passes through the medium, increasing the chances it will stimulate emission of another photon. With enough passes, laser threshold is reached and the laser turns on. Within that limit, one of the mirrors is not perfectly reflecting so that a small fraction of the internal beam is transmitted out of the medium, producing in a typical case a few milliwatts of a highly coherent[6] laser beam.

The kinds of lasers and their applications are too vast to even begin to describe, though lasers all operate on the same basic principle illustrated by the HeNe laser. The two "lasing" energy levels can be in a variety of media, including a gas as with HeNe, a small diode, or a crystal such as

---

[5]The labeling of the energy levels is shown but is not necessary to explain for our limited purpose.

[6]There are also other laser transitions in a HeNe laser at different wavelengths, but operating on the same principles.

the NdYag crystal used in the gravitational wave detector described in Chapter 8. Because of some or all of their beam characteristics, including narrow beams, coherence (see the next paragraph), and high power, lasers are used for a huge number of applications, including holography, drilling, remote sensing, surveying, ranging, communication, guidance, etc.

A fascinating property of a laser beam is its *coherence*, often measured by the *coherence length*, which is the maximum distance along the beam where the front of the beam can show an interference pattern with the back of the beam. Put another way, if the beam is split into two parts following different paths and then rejoined, as in the Michelson interferometer of Chapter 8, the coherence length is the maximum difference in distance the two parts can travel and still show strong interference with each other. In a laser, stimulated emission makes all atoms along the multi-reflected beam path radiate in phase with one another leading to long coherence lengths, from 20 cm to over 20 km. For comparison, in an ordinary gas filled lamp the atoms radiate independently of each other leading to much shorter coherence lengths ranging from a cm down to much less than a mm. Most of the early interference experiments described in this book involved sources with very short coherence lengths, but the differences in path were still small enough to yield strong interference, such as the thickness of a soap film in the first experiment by Young in Chapter 6 or the air-film thickness in Newton's rings.

In this presentation of the HeNe laser above, the Bohr model gives a picture, albeit crude, of energy levels and orbits to help visualize the process. To go further we need a theory that was only suggested by the Bohr model. The new theory was finally found in the mid-1920s and it was called quantum mechanics, a far more radical change from classical ideas than even the Bohr model. There were two separate theories put forth, each of which gave the right answers, the first by Werner Heisenberg, called matrix mechanics, the other by Erwin Schrödinger, called wave mechanics. They were shown to be equivalent theories by Paul Dirac, who later made more innovations in the field. To describe these advances would take us too far beyond our subject of light. We would like, however, to get a glimpse of how quantum mechanics endows material particles such as electrons with wave-like properties, too, so that both light and matter have a wave and a particle nature. Therefore, we present a brief narrative of how the wave nature of material particles was first conjectured in the following short subsection.

The Wave Nature of Material Particles

Wave mechanics got started by Louis de Broglie in 1924 using familiar analogies from optics. We mentioned de Broglie already in Chapter 7, when we pointed out that Fresnel was born on the estate which was Louis de Broglie's ancestral home. De Broglie had an insight contained in his PhD dissertation at the University of Paris, which we will now explain.

De Broglie sought some reason for the quantization condition Eq. (4) that was used by Bohr. It occurred to him that if an electron behaved like a wave, then perhaps the allowed orbits in an atom were just those in which the circumference of the orbit equals an integral number of electron wavelengths. In orbits that did not meet this criterion, the wave would tend to cancel on successive circuits. For circular orbits, he envisaged something like Fig. 14(a), where the electron wave would exist on an orbit of radius r if its wavelength λ fit an integral number of times around the circumference $2\pi r$ of the circle, or $n\lambda = 2\pi r$. All orbits which did not satisfy this condition, as in Fig. 14(b), would not be a stable state. Since $\lambda = h/p$ for an electron[7],

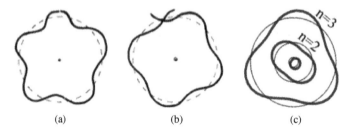

(a)   (b)   (c)

Fig. 14.   De Broglie's explanation of the quantized orbits in atoms using the idea of an electron having particle and wave properties like a photon does. An orbit is allowed if its circumference is an integral number of electron wavelengths as in (a). If it is not, as in (b), the orbit is not allowed. This idea yields the correct orbits for the hydrogen atom as depicted for the lowest three orbits in (c).

(Continued)

---

[7] $\lambda = h/p$ is the same relation that was used for photons in explaining the Compton effect mentioned earlier in this chapter.

(Continued)

this led to the same condition as Bohr's in Eq. (4) for stable states, shown for the first three states in Fig. 14(c). This was the birth of what came to be called wave mechanics. It started with de Broglie's postulate that an electron orbit was a reinforced wave that otherwise moved in that orbit like a particle in a Coulomb field. Then Schrödinger developed his wave equation that described everything in terms of a wave, and the square of the amplitude of the wave at any point gives the probability of finding the particle at that point. (We shall explore this issue of probability as it concerns *photons* in due course.) The wave nature of electrons was confirmed directly by an experiment performed by Davisson and Germer, who scattered a beam of electrons from a crystal. They saw interference between the portion of the beam reflecting from one crystal plane and the portion reflecting from an adjacent plane, similar to the interference between light reflections from the surfaces of a thin film as in Chapter 6.

So, both light and material particles have wave properties and particle properties. At first sight this wave-particle duality seems contradictory. We will discuss this seeming contradiction for the case of light and how it is resolved — the resolution is similar for material particles.

We will find that we must deal with probabilities and uncertainties. First we consider an experiment with a simple beam splitter as shown in Fig. 15(a). A glass plate is covered with a thin film of silver so that it splits a beam of light of a single frequency into two beams: a reflected one and transmitted one falling on separate screens. For convenience we suppose that half the beam is reflected and half is transmitted. Naturally, the screens look equally bright, at least for beams of ordinary brightness. Now we want to look at what happens to individual photons in the beam. For this purpose we reduce the strength of the incident beam so that only one photon passes through the apparatus at a time. The photon must be detected on one screen or the other; if it is detected on the upper screen it will not be found on the transmission screen and vice versa. *There is no way to predict which screen it will be detected on. That is an example of the uncertainty that is central to quantum mechanics.* Of course, the probabilities of being detected on either screen are equal, so that after a large

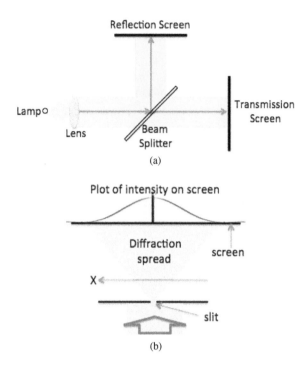

Fig. 15.   (a) A lamp produces light of a single frequency aimed at a beam splitter, which for ordinary intensity reflects 50% of the incident beam and transmits 50%. If the beam is made so feeble that only one photon is incident on the mirror at a time, that photon is detected on either the reflection screen or the transmission screen. We cannot predict in advance which screen. We can just give the probability — 50% for each screen. The quantum theory says this is the best we can do. (b) A photon passes through a slit and creates a diffraction pattern on the screen. A plot of the pattern is shown. The photon is confined to the width of the slit when passing through it, which causes a spread in its X component of velocity due to diffraction. If the slit is made narrower, the spread in velocity component is made larger.

number of photons have passed through, roughly half will be detected on each screen.

Note that detecting a photon, say by ejecting a photoelectron from a spot on the screen, in general destroys the photon. The question then arises: can we measure where a photon is without destroying it? In a way we can, by picking one that has passed through a narrow slit so that we know within the slit width where it is when passing through. How do we know that it has passed through? By detecting it after the slit. However, when we do this, the photon beam forms a single slit diffraction pattern as shown in Fig. 15b (and as we saw in Figure 6 of Chapter 7) which means it has acquired possible motion along the X axis in the figure, the same axis along which it is confined by the slit, with an uncertainty determined by the angular width of the diffraction pattern. In fact, if we narrow the slit further to determine the location better, we spread the subsequent beam further by diffraction, making its motion even more uncertain, since the photon could be anywhere within the new, broader diffraction pattern. This idea of reducing the uncertainty in measuring one variable only to cause an increased uncertainty in measuring another variable has a place of honor in quantum mechanics, and for material particles it is occupied by Heisenberg's famous Uncertainty Principle.[8]

We quantify this experiment by using the variation in wave intensity displayed above the screen in Figure 15(b), which is the single-slit diffraction pattern already shown in Figure 7 of Chapter 7. The pattern is interpreted as giving the relative *probability* that the photon lands at a given point on the screen. If we accumulate a lot of photons, they will fill out the pattern on the screen, with deviations that get fractionally smaller the larger the number of photons. This becomes the classical limit, when the wave picture tells us everything. We will have more to say about the probability interpretation in our third example given next.

Now we consider what happens when wave interference occurs, as we have treated many times in this book, but this time combined with the particle notion of irreducible quanta $h\nu$. We will use the well-known two-slit interference experiment we discussed in Chapter 7 to highlight the issues and see how they are resolved. We will find something amazing at first: each photon *interferes with itself* to produce the interference pattern. In the two-slit experiment, recall that a diffraction-limited beam of light

---

[8]This Principle states that the position x and momentum p of a material particle cannot be measured along the same axis to unlimited precision at the same time. The product of the uncertainties $\Delta x \Delta p$ is about h or larger. Using $p = h/\lambda$ found in the Compton Effect for a photon, the analysis of diffraction suggests a similar result along the X axis for a photon, but there are difficulties defining the position of a photon. For the learned reader, see T. D. Newton and E. P. Wigner, *Rev. Mod. Phys.* **21**, 400 (1949).

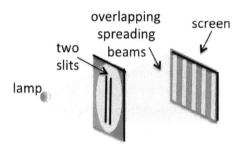

Fig. 16.   From Figure 3 of Chapter 7: A diffraction-limited beam from a lamp spreads out to illuminate two slits from which two beams spread out and overlap on a screen, thereby forming an interference pattern of alternate light and dark fringes as shown. What happens when the beam in reduced until only one photon is in the apparatus at a time? The answer is given in the text and in Fig. 17.

passes through two slits, forming two beams which then fall on a screen, creating the by now familiar interference pattern shown in Fig. 16 (taken from Figure 3, Chapter 7). This pattern consists of alternate bright and dark bands or fringes, where the two beams are either in phase with each other forming a bright fringe or out of phase, canceling each other out and forming a dark fringe. That is what we get when we calculate the distribution of light on the screen using the wave theory, and what we see if the illumination is sufficient to represent a large number of photons during a measurement time. For even weak illumination the number of photons is huge; a nanowatt of visible light (10^{-9} watts) still contains over a billion photons per second.

Let's now examine what happens if we turn down the illumination so low that, for example, only one photon lands on the screen at a time, much as we did with the beam splitter. When a photon is passing through the slits, no other photon is around. This reduced illumination forms the same calculated interference pattern on the screen — it is just much weaker. Just as with the single slit, the *probability* that the photon will land at any one place is proportional to the illumination there. In Fig. 17 we can observe the interference pattern build up over time as more photons land. Note especially that no photon ever lands at a place that will become a dark fringe when a lot of photons have landed, the

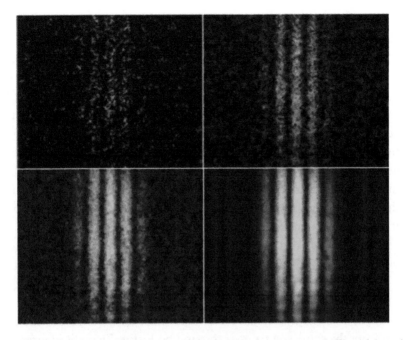

Fig. 17.   Photons passing through two slits and landing on a screen as in Fig. 16, but with the brightness of the source (a green laser pointer) mostly blocked with attenuating filters, so that we can see individual photons arrive. The exposure time is increased from left to right. The probability a photon will land at a place on the screen is proportional to the intensity of light at that place. The pictures show a two-slit interference pattern building up. Each photon passes through both slits and interferes only with itself.

Source: Dimitrova T. L. and Weis A. *Am. J. Phys.*, **76** (2008) 137–142.

most striking proof that each photon "knows" both slits are there. *Each photon must travel through both slits and interfere with itself* to produce the pattern on the screen. This is another idea that is difficult to recon-cile with "common sense". Until the photon is detected, it could show up at any place the pattern has a finite value. When it is detected, the game is up. The photon is at the one place where a spot shows up on the screen in Fig. 17, and the entire wave for that photon exists no more; as some say, the wave has collapsed at the point where the photon is observed. To emphasize once again, the probability a photon will land at a place on the screen is proportional to the relative intensity of the interference pattern at that place, and then it shuts off; that photon cannot show up anywhere else once observed.

Although it may seem weird to have one photon going through both slits and interfering with itself, a similar process for all the interference examples we have discussed in this book. Our first example was thin film interference back in Chapter 6, Figure 3, in which one beam produces two reflected beams: one from the front and one from the rear of a soap film, which then interfere. But many individual photons in the reflected beam have taken two paths: (1) reflecting from the front face of the film and (2) being transmitted and then reflecting off the back face. It is these two paths for the <u>same</u> photon that causes the interference. One might think that a photon is either reflected or transmitted from the front face, but the interference shows that each photon does both, and then interferes with itself. The same is true for the partially reflecting mirror in the interferometer in Chapter 8, Figure 9. So if one adopts the view that each photon interferes with itself, all the interference phenomena we have studied are explained; two-slit interference can take place only when the incident beam from the same source goes through both slits. Two-slit interference has received a lot of attention over the years, and still does, especially the "strange" fact that each photon goes through both slits. However, it is perhaps just as strange that each photon samples both path 1 and path 2 of a thin film to produce interference in that case as well. (None of these examples seems so strange when you get used to it.)

We can summarize this discussion with a simple set of rules. When a photon is emitted it has an associated wave; the behavior of the wave is to be calculated by the normal rules for waves we have developed for each situation, including all diffraction and interference effects, leading to some illumination pattern on a screen or detector. The relative probability of the photon being detected at any point is proportional to the intensity of the light at that point, which results from all pathways the photon takes to get there. Before it is detected, there is no way to predict with certainty where it will land, just with relative probabilities. The photon disappears once it has been detected.

At its inception, many people didn't like this intrinsic uncertainty of quantum mechanics and postulated so-called *hidden variables*, which if we only knew and could control, would make the randomness go away and we would have a deterministic universe. In the two-slit experiment, the variable would determine with certainty where a photon would land on the screen. So far, no such variables have been found, and most physicists believe there are none.

So we have come to the end of our story and find that its subject, the nature of light, is in general one of probabilities, not certainties. This may be a surprising outcome for those who expected certainty and determinism. Some may still not believe this outcome. For readers who share that view, you are in good company: Einstein could never accept the uncertainty of quantum mechanics. But for many years now, the overwhelming majority of physicists accept the uncertainty we have described. The nature of light is both wave and particle. In the case where this nature causes uncertainties, we can actually calculate what the uncertainties are and calculate the average result with great accuracy.

Summary of Advanced Developments

First there is quantum electrodynamics, or QED, which started with the notion we have already learned that light (and in general the electromagnetic field) consists of quanta of energy $h\nu$, which we call photons. Beginning about 1930, physicists began using such quantum notions to calculate how a charged particle such as an electron could interact with its own radiation field. As one example, an electron in a bound state of atomic hydrogen could emit and reabsorb the same photon, staying in the same atomic state but causing a small shift in the energy of the state. Such energy shifts are called radiative corrections and occur for all atomic states. By the late 1940s theorists could actually calculate this shift in energy for the $n = 2$ state of atomic hydrogen, a shift which had just been measured, and theory and experiment agreed to high accuracy! So QED became a very successful theory and a model to extend to the other known interactions; namely (1) the so-called weak interaction in which neutrinos at then attainable energies have a feeble coupling to other particles, (2) the strong interactions, the force that binds nucleons together to form nuclei, and (3) gravity. In the late 1960s electroweak unification was finally achieved, that is QED and the weak force were

---

(*Continued*)

(Continued)

unified into one theory, with new predictions that have all been verified. Attempts at further unification to include the strong interaction as well have been largely successful. Meanwhile, gravity still stands apart, but with ever-more sensitive detection of gravitational waves by LIGO as we discussed in Chapter 8, and other experiments, certainly more will be learned. As for now, however, unification of gravity with the other forces remains a grand dream.

Another area of advances has been the application of the complete quantum theory to light, a field sometimes called quantum optics. Currently active topics have such suggestive names as quantum entanglement, quantum teleportation, and quantum logic gates. The latter are of much interest in quantum information theory and ultimately could be used in quantum computers. However, light-based technologies are only one of many competing technologies aimed at realizing this goal.

Before concluding we should mention the remarkable application of lasers in the past 40 years to trapping atoms in light beams and to laser cooling of trapped atoms and atomic ions to unprecedented low temperatures. In trapped gases at such low temperatures, new states of matter have been created. Other experiments have produced the most precise atomic clocks yet, and still others continue to enable the studies of quantum optics mentioned in the previous paragraph. We cannot explain such advances at the level of this book, but we can emphasize their importance by noting that since 1989, six years of Nobel prizes in physics have honored at least one of the recipients each year for work in laser cooling and trapping.

To get to this conclusion, our journey has been a remarkable one; it has taken us through the discovery of many wonderful properties of light, both wave and particle, and it is clear that applications of these properties with improved instruments will continue in directions we cannot foresee now. Has the understanding of the nature of light progressed further than what we have been able to present at the level of this book? The answer is yes, and we now offer a brief summary of these advances in the above subsection.

The study of light has certainly come a long way from when Descartes computed the size of the rainbow and showed that Snell's Law of refraction is correct. The journey since then has been filled with stories of adventure and of mysteries solved. I hope the reader has enjoyed the journey and also gained an appreciation of the beauty of how all the parts revealed in the journey fit together in one theory of waves and particles. This one theory continues to have surprises, however, which are being found in current research, particularly in quantum optics. Thus, we can be sure that discoveries still await us at the frontiers of the field, and though these discoveries *seem* unlikely to undermine the basic tenets of the field they will continue to add to our understanding and appreciation of the nature of light.

Index